T0221126

Carbon

Resources series

Carbon

KATE ERVINE

polity

First published in 2018 by Polity Press

Polity Press
65 Bridge Street
Cambridge CB2 1UR, UK

Polity Press
101 Station Landing
Suite 300
Medford, MA 02155, USA

ISBN-13: 978-1-5095-0111-3
ISBN-13: 978-1-5095-0112-0 (pb)

A catalogue record for this book is available from the British Library.

Library of Congress Cataloging-in-Publication Data
Names: Ervine, Kate, author.
Title: Carbon / Kate Ervine.
Description: Cambridge, UK : Polity Press, 2018. | Includes bibliographical references and index.
Identifiers: LCCN 2018004919 (print) | LCCN 2018013626 (ebook) | ISBN 9781509501151 (Epub) | ISBN 9781509501113 (hardback) | ISBN 9781509501120 (pbk.)
Subjects: LCSH: Carbon--Environmental aspects. | Carbon dioxide mitigation. | Carbon offsetting. | Carbon offsetting--Political aspects. | Emissions trading--Political aspects.
Classification: LCC QD181.C1 (ebook) | LCC QD181.C1 E78 2018 (print) | DDC 363.738/74--dc23
LC record available at https://lccn.loc.gov/2018004919

Typeset in 10.5 on 13pt Scala by
Servis Filmsetting Ltd, Stockport, Cheshire
Printed and bound in the UK by CPI Group (UK) Ltd, Croydon, CR0 4YY

For further information on Polity, visit our website: politybooks.com

For Sasha, Sebastian, and Gavin

Contents

Figures and Tables

Acknowledgements

Writing a book is never an independent endeavour and much is owed to those who have helped in the process. Many thanks to Amr El-Alfy and Janet Music for the excellent research assistance they provided, to two anonymous reviewers for their insightful feedback, and to Gavin Fridell for taking the time to read and comment on the manuscript. I am grateful to Louise Knight and Nekane Tanaka Galdos at Polity Press for their superb and patient editorial assistance, and to the Social Sciences and Humanities Research Council of Canada and Saint Mary's University for supporting various aspects of my research. I owe an immense debt to the many individuals who have graciously offered their time as participants in my research on emissions trading and global carbon markets and which has helped to shape, in so many ways, how I think about carbon. To those organizations and groups here in Nova Scotia that are working tirelessly on issues of carbon, climate change, and justice, including the Affordable Energy Coalition, the Canadian Centre for Policy Alternatives, and the Ecology Action Centre, an immense thank you for refusing to accept business as usual, and for allowing me to join you in your work. On a more personal note, I am thankful for the friendship and scholarly support I have received from Claudia De Fuentes, Aldona Wiacek, and Lyuba Zhyznomirska; our collective writing strategies were instrumental in helping to get the job done. I remain

Acknowledgements

deeply grateful to my parents, Bernadette and Harold, for their support and encouragement in all that I do, and to my sister Shawna for her friendship and for all the laughs. Most importantly, I want to thank Gavin for his unwavering support and encouragement every step of the way, and Sasha and Sebastian, for the love and joy they bring to life. As I watch them grow, I am constantly reminded of the intergenerational threat that is catastrophic climate change. It is for them and all those who deserve a better world that I write this book. Any omissions or errors are my responsibility alone.

Halifax, 2018

Abbreviations

AAU	Assigned Amount Unit
AOSIS	Alliance of Small Island States
AR	afforestation and reforestation
BAU	business as usual
BECCS	bioenergy with carbon capture and storage
CAA	Clean Air Act
CBDR-RC	common but differentiated responsibilities and respective capabilities
CCS	carbon capture and storage
CDM	Clean Development Mechanism
CDP	Carbon Disclosure Project
CER	Certified Emission Reduction
CH_4	methane
CO_2	carbon dioxide
COMFIT	Community Feed-in Tariff Program
COP	conference of the parties
CORSIA	Carbon Offset Reduction Scheme for International Aviation
CPLC	Carbon Pricing Leadership Coalition
CSR	corporate social responsibility
DAC	direct air capture
EDF	Environmental Defense Fund
EJ	environmental justice
ENGO	environmental non-governmental organization
EOR	enhanced oil recovery
EPA	Environmental Protection Agency

ERU	Emissions Reduction Unit
ETS	Emissions Trading System
EU	European Union
EUA	European Union Allowance
EU ETS	European Union Emissions Trading System
EV	electric vehicle
EW	enhanced weathering
FIT	Feed-in Tariff
GCC	Global Climate Coalition
GHG	greenhouse gas
$GtCO_2$	gigatonne of carbon dioxide
$GtCO_2e$	gigatonne of carbon dioxide equivalent
GW	gigawatt
GWP	Global Warming Potential
HFC	hydrofluorocarbon
IAM	Integrated Assessment Model
ICAO	International Civil Aviation Organization
IEA	International Energy Agency
IETA	International Emissions Trading Association
IMF	International Monetary Fund
INDC	Intended Nationally Determined Contribution
IPCC	Intergovernmental Panel on Climate Change
ITMO	internationally transferred mitigation outcome
JI	Joint Implementation
LDC	least developed country
LNG	liquefied natural gas
MRV	measuring, reporting, and verification
N_2O	nitrous oxide
NDC	Nationally Determined Contribution
NET	negative emissions technology
NGO	non-governmental organization
OECD	Organisation for Economic Co-operation and Development
ppm	parts per million

REDD+	Reducing Emissions from Deforestation and Forest Degradation
RGGI	Regional Greenhouse Gas Initiative
SDM	Sustainable Development Mechanism
SIDS	Small Island Developing States
SO_2	sulphur dioxide
SRM	solar radiation management
tCO_2e	tonne of carbon dioxide equivalent
TNC	The Nature Conservancy
TWh	terawatt hour
UN	United Nations
UNDP	United Nations Development Programme
UNEP	United Nations Environment Programme
UNFCCC	United Nations Framework Convention on Climate Change
WCI	Western Climate Initiative
WRI	World Resources Institute
WWF	World Wildlife Fund

CHAPTER ONE

The Problem of Carbon

It is only fairly recently that carbon has become an issue of grave concern globally, a major preoccupation for politicians, activists, citizens, and business leaders worldwide. If in the past many were content to leave the study of carbon to the natural sciences, this is no longer the case, primarily because carbon – or, more specifically, carbon dioxide – is the main heat-trapping gas in our atmosphere and it is dangerously warming our planet. And because of carbon's central role in contributing to global climate change, it has become one of the world's most contested resources. This condition is nothing if not paradoxical; while carbon is absolutely essential to supporting life on Earth, too much of it threatens to destroy life as we know it. This book has been written in part so that we might demystify the paradox. Indeed, if we are to have any hope of understanding how it has come to pass that carbon, this most critical of chemical elements, without which none of us would exist, has been transformed into an element capable of stoking our deepest existential anxieties, we must take the carbon out of 'nature', so to speak, situating it within the broader political, economic, social, and cultural processes that have shaped human history, good and bad, since the onset of the industrial era. Especially central to this history are fossil fuels – coal, oil, and natural gas – whose energetic properties have powered economic growth and development globally since the onset of industrial capitalism over 200 years ago, and whose burning releases carbon dioxide into the atmosphere. At the time of writing, fossil fuels continue to make up over 80 per cent of the world's energy mix, responsible for over 90 per cent of global carbon dioxide emissions between 2000 and 2011. These figures hint at why dealing with global climate change has become so utterly intractable; as we will see below, carbon is embedded in the very fabric of contemporary society and its specific brand of politics, presenting us with one of the most complex and far-reaching problems ever to face humanity.

In order to illustrate this, it is fitting to begin a book on carbon with a reflection on the politics of the everyday. By drawing attention to mundane lived experiences, we are better positioned to recognize carbon's ubiquity in so much of what it is we do. For me, today was a day that started like most. After waking to the sound of my electric alarm clock, I got out of bed, flicked on the lights, and took a hot shower powered by our electric water heater. From there, I turned up the oil-generated heat in our house, fed our cat her refrigerated food made with industrial chicken not destined for human consumption, and got myself ready for the day. The other members of my family engaged in a similar ritual, finally meeting downstairs to make breakfast. Without electricity – approximately 70 per cent of which comes from fossil fuels in my home province of Nova Scotia, Canada, with 55 per cent of that coming from coal, the most carbon-intensive of all fossil fuels – our fridge and blender would have been useless, while, without fossil-fuelled long-distance trade, shipping, and industrial agriculture, the bananas from Ecuador, apples from the United States, and hemp seeds from Manitoba would not have been available, especially in the middle of the Canadian winter. Having finished all of this, the kids went off to school and daycare with their dad in the family's gas-powered car. I now sit at my laptop computer – one of two in the house – drinking coffee from Nicaragua, beginning a day of work from home. The pathway that eventually got me here included my time as a university student in the Canadian province of Ontario, when I was lucky enough to land a coveted high-paying job for three summers at General Motors, the auto manufacturer that employed my father for some forty years. As a graduate student, I was even luckier in securing a well-paid job for two summers in a steel mill. These jobs, in two of the most carbon-intensive industries globally, were critical sources of income to pay my tuition, my relationship with General Motors extending back to the day I was born and the income it provided to my family to buy the food, clothes, and shelter necessary for survival. Without a doubt – and whether I like it or not – my fossil-fuelled life, past, present, and future, is intensely carbonized, and it is certainly not unique.

Global society today is more fully fossil-fuelled and carbonized than at any point in history, with carbon emissions in 2017 reaching their highest level ever. While inequalities of power, class, race, gender, and geography mean that the extent of fossil fuel use and subsequent carbon emissions vary dramatically between individuals and countries, most of us are, nevertheless, deeply embedded in complex carbon networks and dependencies that are structured according to historically rooted patterns of growth and development. Dealing with the threat of too much carbon becomes especially difficult in this context. On the one hand, advanced consumer societies carry with them a particular cultural politics whose roots reach back to the advent of industrial capitalism, a system that emerged in tandem with stunning advances in science and technology. Though the expansion of this system has depended deeply on the exploitation and oppression of so many of the world's peoples and environments, it has, nevertheless, always carried with it the promise of emancipation. In this, the grand narratives that continue to define what remains a largely fossil-fuelled and carbon-intensive industrial capitalism are rooted in powerful notions of the 'good life' – modernity, linear progress, security, and comfort. There is much that we as individuals have invested in sustaining these civilizational narratives, for acknowledging their cracks strikes at the very heart of whom we tell ourselves we are, and what we hope for ourselves and for our futures. On the other hand, addressing the problem of too much carbon is significantly complicated by the very basic necessities of biological and social reproduction. Where does one work to feed, clothe, and shelter their family, or to pay for school? How does one heat their home, keep the lights on, and cook their food? In this, the life opportunities of many, worldwide, remain dependent upon carbon-intensive modes of production and income. And while there are many across the globe who do not accept industrial society's grand narratives, and many whose livelihoods remain minimally carbonized, the truth of the matter is that too many of us are fully immersed within and dependent upon carbon in one form or another, with our political systems all too often structured to respond to the interests of those with the greatest stake in maintaining business as usual.

According to scholars publishing in the prestigious science journal *Nature* in 2015, in order to have a 50/50 chance of limiting global temperature increase to below 2 °C above pre-industrial levels – the politically agreed, though insufficient, limit past which dangerous and catastrophic climate change is all but assured – 88 per cent of global coal reserves, 52 per cent of global gas reserves, and 35 per cent of global oil reserves cannot be burned. If you find 50/50 odds highly unsettling when gambling on the future of our planet, you would probably agree that those percentages demand a significant upward revision. Yet, according to estimates from the International Energy Agency (IEA), global fossil fuel subsidies on a pre-tax basis were US$493 billion in 2014. When the International Monetary Fund (IMF) calculated these subsidies on a post-tax basis to account for the social and environmental harm associated with fossil fuel use, they ballooned to over $5 trillion. Meanwhile, global subsidies to the renewable energy sector in 2014 were estimated at $120 billion, four times less than pre-tax subsidies for planet-warming fossil fuels. In spite of the scientific data predicting that we could be on course to a disastrous temperature increase of 3–6 °C above pre-industrial levels, hundreds of billions are spent every year to find, extract, and burn more fossil fuels than we can ever use, with trillions more spent to deal with the devastating environmental and social costs of our fossil fuel addiction. Given that over 90 per cent of global carbon dioxide emissions come from fossil fuel combustion and cement production, with land use change, including agriculture and deforestation, responsible for the remaining 10 per cent, a book about carbon is necessarily a book about fossil fuels due to their disproportionate contribution to global climate change.[1]

While it is tempting to reduce the problem of transitioning away from fossil fuels to a technical issue – one of simply replacing carbon-intensive energy sources with non-carbon energy sources – the carbonized networks, structures, and dependencies discussed above are shaped profoundly by very specific relations of power and wealth globally; a transition away from these networks, structures, and dependencies represents, thus, a serious threat to those who disproportionately benefit and profit from them. This is what makes carbon, in part, so deeply contested and conflictual. While fossil fuel companies are some of the most obvious beneficiaries of this system – recent research reveals that a mere twenty-five of these companies are responsible for half of all carbon emissions since 1988 – the benefits extend well beyond specific companies. When measured on a per capita basis, World Bank data shows that, between 2011 and 2015, the average Canadian citizen emitted 14.1 metric tons of carbon dioxide (CO_2) per year, the average American 17.0, and the average German 8.9. This compared to the average Chinese citizen's per capita emissions of 6.7, the average Indian's of 1.7, and the average Sierra Leonean of 0.2.[2] While per capita statistics fail to tell us who is emitting the most in any particular country, and we know that wealthy sectors of the population have higher carbon footprints than their lower-income and poor counterparts, this data hints at how carbon is implicated in global patterns of wealth and inequality. One of the most contentious issues that has complicated global efforts to address the problem of too much carbon has to do with the fact that it was precisely through carbon-intensive fossil-fuelled growth that the global North was able to achieve its advanced levels of wealth and development. And, in so many instances, this wealth and development depended on colonizing nations and peoples around the world, devastating their populations and limiting their ability to replicate the carbon-intensive path pursued by the North. This history, within which the few have consumed and polluted far beyond their fair share, simultaneously leaves the world's poor and marginalized, those least responsible for climate change, bearing the devastating brunt of its impacts. There is a compelling case to be made that a book about carbon must also be a book about the quest for social and environmental justice and democracy, given the role carbon has played and is playing in our profoundly unequal world.

In the pages that follow, the story of carbon will unfold. Starting with the carbon atom, ubiquitous in nature and without which life on Earth would not be possible, we'll examine how the planet's long-term carbon cycle, largely stable for hundreds of millions of years, began to experience a dramatic destabilization some 200 years ago, with concentrations of CO_2 beginning their dramatic ascent to their present levels. From there, we will consider the perilous path that is unfolding with too much carbon in the atmosphere, examining its impacts in the here and now, along with the deep threats it poses to the future of human life on this planet. All of this will prepare the groundwork for chapter 2 and its focus on how we got here, and why it remains so difficult to kick the carbon habit.

Carbon as life

Carbon – C in the periodic table – is a chemical element abundant in the world around us. In its purest state, it is the diamond on one's finger or the graphite in your pencil, but, once joined to other elements, it forms carbon compounds, of which there are millions. Carbon and its compounds are essential building blocks supporting all forms of life on Earth, but when it comes to this most critical of elements, you can have too much of a good thing. The carbon that forms the backdrop to this book, and whose increasing abundance in the earth's atmosphere has already begun to warm the planet in highly destructive ways, is carbon dioxide, or CO_2, a gaseous compound forged when carbon and oxygen atoms bond. Before getting to CO_2 as greenhouse gas, however, it is useful to review why we cannot live without it.

All life forms on Earth need carbon's energy to sustain that life, and to grow and thrive. We can trace carbon's path to us as humans by starting with photosynthesis, the process through which the sun's solar energy is captured and used by plants, algae, and some bacteria to convert atmospheric CO_2 and water into carbon compounds such as glucose, starch, and cellulose, used in the development of these species. These autotrophs, so named because they produce their own energy through photosynthesis, then serve as the main source of food energy for heterotrophs – humans and animals – who, unable to self-manufacture their food, must consume organic substances, either plant- or animal-based. Once consumed, these carbon compounds, including proteins, carbohydrates, and fats, comprise the building blocks through which heterotrophs grow and develop. Through cellular respiration, whereby cells break down carbon compounds to produce the energy that supports life functions, CO_2 is generated as a waste product, and released back into the atmosphere. While highly simplified, this description of carbon cycling from photosynthesis through to cell respiration, and within which CO_2 is removed from and then released back into the atmosphere, illustrates the process through which CO_2 cycles through the Earth's biosphere. Also included in the biospheric carbon cycle are processes through which carbon compounds make their way into soils through excretion, death, and decomposition, and are broken down by microbes, with CO_2 released as a by-product of soil respiration; and the process through which CO_2 is dissolved in oceans, to be used by phytoplankton in photosynthesis to produce carbon compounds to aid in the development of ocean life, with CO_2 released through respiration.

Through these processes, carbon serves as a foundational element supporting the internal growth and development of all planetary life forms. It was some 3.5 billion years ago, in fact, that early forms of photosynthesis began with single-celled molecules, and it was between 2.7 and 2.8 billion years ago that multi-cellular organisms began to produce oxygen as a by-product of photosynthesis. Indeed, by analysing the planet's geological history, we now understand the critical role carbon has played in the emergence and evolution of life on Earth, with the appearance of our earliest ancestors some 6 million years ago requiring a sufficiently oxygenated atmosphere, impossible without carbon, to support our presence. Humans and their ancestors simultaneously required planetary temperatures able to support the life forms unique to Earth. CO_2 possesses heat-trapping properties – hence the name "greenhouse gas" – that trap a portion of the sun's infrared radiation within the Earth's atmosphere, heating the planet sufficiently to support the abundant planetary life forms that we know today.[3]

In short, carbon is everywhere we look and in everything we touch. For our bodies, our food, our plants, our oxygen, our temperatures, and so much more, we can thank carbon. It is the carbon flowing through the myriad pathways and networks from one site to the next that forms, in turn, the global carbon cycle.[4] According to the Intergovernmental Panel on Climate Change (IPCC), the global carbon cycle comprises 'a series of reservoirs of carbon in the Earth System, which are connected by exchange fluxes of carbon'. Scientists separate the global carbon cycle into two dominant and interacting domains, one fast and the other slow, according to the speed at which carbon flows through these reservoirs. Reservoirs in the fast domain, through which carbon flows at speeds of seconds to thousands of years, and within which human beings actively participate, include Earth's atmosphere, vegetation, fresh water, soils, and oceans, including sea-floor sediments. Carbon flowing through the slow domain, taking upwards of 10,000 to millions of years, is concentrated in Earth's geosphere – its rocks and sediments – interacting with the fast domain through 'volcanic emissions of CO_2, chemical weathering, erosion, and sediment formation on the sea floor'. The IPCC notes the exchanges between the slow and fast domain have remained largely stable for the last few centuries, with a similar 'steady state' in exchanges within the fast domain for the vast majority of the last 11,700 years, the geologic epoch known as the Holocene.

In fact, during the Holocene, natural fluxes in the global carbon cycle have seen Earth's oceans and terrestrial biota serve as net carbon sinks, absorbing more carbon annually than has been released back into the atmosphere. Beginning around 1750 with the onset of industrialization, however, concentrations of atmospheric CO_2 began to increase significantly, disrupting the relative stability of annual carbon exchanges. In particular, the IPCC notes that concentrations of atmospheric CO_2 are now higher than at any time over the past 800,000 years, increasing from 278 parts per million (ppm) in 1750 to 390.5 ppm in 2011, and representing a 40 per cent increase above pre-industrial times. In late May 2016, concentrations of CO_2 in the atmosphere were measured at 407 ppm. As noted above, this perturbation in the global carbon cycle is largely due to the combustion of fossil fuels, and to a lesser extent to changes in land use, both anthropogenic in origin, and leading to a significant imbalance whereby more CO_2 is released to the atmosphere annually than can be absorbed by existing reservoirs.

Fossil fuels, or hydrocarbons, were formed over hundreds of millions of years in a process that saw prehistoric plants and animals decay, to be covered by layers of sediment, mud, rocks, and sand. As the layers of organic matter built up in conjunction with the ebbs and flows of the planet's long geological cycles, saturated environments deprived of oxygen were subject to increased pressure and high temperatures, creating the ideal conditions for the eventual formation of fossil fuels – coal, oil, and natural gas. As carbon compounds composed of hydrogen and carbon, hydrocarbons in the form of fossil fuels represent a significant carbon reservoir within the Earth System, but one that largely sequesters carbon so that it does not play a role in the natural carbon cycle described above. It is only when humans extract and burn fossil fuels, of which CO_2 is the dominant by-product, that these ancient stores of carbon reenter the carbon cycle, massively disrupting its relative steady-state achieved during the Holocene, which has been crucial to supporting planetary conditions suitable for the development of humankind.

Carbon dioxide as greenhouse gas

Without our atmosphere, Earth would be uninhabitable in its present form. Composed of five main layers, Earth's atmosphere serves multiple functions that help to sustain the planetary ecosystem. These include blocking out harmful ultraviolet shortwave radiation from the sun; limiting the amount of shortwave solar radiation that reaches the Earth's surface so that the planet does not overheat; maintaining an oxygenated environment; and serving as a planetary thermostat by regulating the outflow of longwave infrared radiation so that a sufficient amount of thermal energy stays on Earth. When it comes to temperature, a number of processes are worth highlighting. First, the sun's shortwave solar radiation, largely visible as light, reaches Earth, with just under a third reflected back to space by the atmosphere, and the remainder passing through. When it reaches the surface of the Earth, this solar energy is converted into heat energy warming the planet, with longwave infrared, or thermal, radiation reemitted. In turn – and this is critical to our understanding of anthropogenic global warming – greenhouse gases residing in the lowest layer of the atmosphere, known as the troposphere, absorb thermal infrared radiation and reemit it either back to Earth or out to space. A full 99% of the gases that make up the atmosphere are non-greenhouse gases, including nitrogen (N_2) at just over 78%, and oxygen (O_2) at just over 20%. Of the greenhouse gases (GHGs) that make up the remaining 1%, carbon dioxide is the most plentiful at approximately 0.04%, followed by methane (CH_4) at around 0.000179%, and nitrous oxide (N_2O) at 0.0000325%. Additional trace GHGs include water vapor (H_2O), ozone (O_3), and chloro-fluorocarbons (CFCs).

It is the molecular structure of GHGs that help to explain their heat-trapping properties. Unlike nitrogen and oxygen molecules, each composed of two atoms and possessing an equal number of positive protons and negative electrons that cancel each other out, GHG molecules such as carbon dioxide, nitrous oxide, and methane contain three or more atoms. Their asymmetry, or lopsidedness, means that GHG molecules vibrate in such a way that their electrical fields change, allowing them to absorb infrared light, or heat. Put simply, following the conversion of the sun's incoming shortwave solar radiation to infrared radiation, GHGs in the atmosphere absorb a portion of the outgoing infrared radiation, which is then reemitted either back to Earth or out to space. With the overall outgoing infrared radiation reduced, the planet is warmed, with GHGs functioning in a manner similar to a greenhouse.[5]

As the most abundant GHG in our atmosphere, CO_2 is critical to supporting the evolution of life on Earth; without it, Earth's distance from the sun would result in a largely frozen landscape. According to the IPCC, evidence taken from ice core measurements reveal that for 800,000 years, until 1750, concentrations of atmospheric CO_2 were in the range of 180 ppm to 300 ppm, corresponding to the glacial and interglacial periods. Represented differently, CO_2 made up approximately 0.018 to 0.03 per cent of the atmosphere during this time. As noted above, it was in 1750 with the onset of the industrial era and when atmospheric concentrations of CO_2 stood at 278 ppm, that quantities of CO_2 began their increase to arrive at over 400 ppm, and they keep climbing. While monthly CO_2 concentrations display seasonal variation linked to the role of vegetation die-off and regrowth in the global carbon cycle, research published in June 2016 in the journal *Nature Climate Change* predicted that we would never again in 'our lifetimes' see monthly CO_2 concentrations below 400 ppm, pointing out the 'iconic milestone' this represented in the steady march of anthropogenic climate change.[6]

Moreover, these findings help to underscore another pernicious problem with CO_2, related to its resident time in the atmosphere once emitted. Scientists researching the lifetime of fossil fuel CO_2 in the atmosphere estimate that 60–80 per cent will transfer out of the atmospheric carbon reservoir within 200–2,000 years, with the remaining 20–40 per cent staying in the atmosphere for tens to potentially hundreds of thousands of years, as it awaits much slower chemical reactions with Earth's geosphere.[7] Their focus on fossil fuel CO_2 is important – between 1750 and 2011, fossil fuel CO_2 accounted for approximately 65 per cent of anthropogenic CO_2 emissions globally, followed by roughly 32 per cent from land use change, and the remainder coming from calcination associated with global cement production. More importantly, and reflecting the recent rapid industrialization in the world's emerging economies in addition to persistent high emission rates in the world's advanced industrial economies, IPCC data shows, as already discussed, that from 2000 to 2011 CO_2 emissions from fossil fuel combustion accounted for close to 90 per cent of global emissions, and were approximately 54 per cent higher than they were in 1990.

In short, given that CO_2 emissions are cumulative, we know that at least some of the CO_2 emitted in the nineteenth century remains with us today; in turn, the science tells us that the hundreds of billions of tonnes of CO_2 emitted since 1750 are dangerously warming the planet. In sum, we are rapidly injecting the atmosphere with colossal quantities of heat-trapping molecules that, once in the door, turn out to be the house guests who had no intention of ever leaving. It is to the task of unpacking the catastrophic consequences of this process that we now turn.

A perilous path

Doing a quick Google search for 'catastrophic climate change' turns up a mere 16.2 million hits compared to the 34.4 million hits when substituted with 'dangerous climate change'. Topping the lists are publications from scientific bodies such as the IPCC and academic research institutes, academic journal articles, large environmental non-governmental organizations (ENGOs), and countless news articles from well-recognized and reputable sites. Following the historic Paris Climate Change Conference in December 2015, hosted by the United Nations Framework Convention on Climate Change (UNFCCC), talk of the perils of climate change seems to be everywhere. But what are those perils? What happens when atmospheric concentrations of CO_2 increase from 397 ppm to 407 ppm, then to 450 and beyond? What happens if the planet experiences a global temperature increase of 2 °C above pre-industrial times, the limit agreed to by global leaders in 2009 and seen at the time as essential to avoiding dangerous global warming? What do we make of research that suggests 2 °C may be the 'threshold between dangerous and *extremely* dangerous, rather than between acceptable and dangerous climate change' (p. 20)?[8] What do things look like at 1.5 °C of warming, the aspirational target in the Paris Agreement in recognition of the fact that, for many, including many Small Island Developing States (SIDS), an increase of 2 °C would ensure their destruction? What is happening now that the planet has already warmed 1 °C above pre-industrial times, another dire milestone announced just before the Paris Climate Change Conference? What are the implications of analyses showing that the GHG reduction pledges made by 189 governments in Paris put us on track for a devastating 3–4 °C of warming – and that is only *if* governments actually fulfil their commitments. What if they do not?

Climate change in the present
If we start in the present, there is no shortage of news stories on any given day showing that, at 1 °C of warming, the planet has entered a climate emergency.[9] Temperature data reveals levels of warming that are considerably more rapid and extreme than in earlier projections from climate scientists, including data showing that 2016 was the hottest year ever recorded, a title snatched from 2015 which itself had beat out 2014 for the position. A full sixteen of the seventeen hottest years ever recorded have occurred since the year 2000. As this chapter was being prepared, extreme heat waves were gripping parts of India, where a record-setting 51 °C was reached in the state of Rajasthan in May 2016, melting roads, killing crops, drying up water

sources, and leading to power outages. This was followed by record temperatures of 48 °C in July 2016 in parts of the United States, fuelling wildfires and drought, which were followed that same month by what are likely world record-setting temperatures of 54 °C in Mitribah, Kuwait. Data shows that in Fujairah, United Arab Emirates, soaring temperatures and extremely high humidity combined to produce 'off the charts' heat index values of 60 °C. In 2015, parts of India underwent a heat wave that killed over 2,300; while Pakistan's 2015 heat wave, responsible for over 1,300 deaths, prompted officials in May 2016 to begin digging mass graves in preparation for a possible repeat as summer approached. The death toll from Russia's 2010 heat wave was a stunning 55,000, resulting in a loss of approximately 25 per cent of its annual crops, wildfires that burned over 1 million hectares, and $15 billion in economic losses. It is estimated that over 70,000 died as a result of Europe's 2003 heat wave, showcasing climate change as an extreme threat to human health and public health systems.

With greater frequency, extreme heat waves occur along-side drought, stressing water and agricultural systems, with significant economic costs as productivity losses mount and energy infrastructure is stretched. California's devas-tating drought, lasting from 2011 to 2016, led to losses of $5.2 billion for the state's economy in 2015, with 21,000 direct and indirect job losses in the agricultural sector. In May 2016, wildfires linked to climate change and abnor-mally dry conditions ripped through parts of the Canadian province of Alberta and its iconic Fort McMurray, famous as the epicentre of Canada's tar sands production, destroy-ing homes, businesses, First Nations territories, and critical infrastructure, with estimates that insurance claims would reach between CAD$4.4 and $9 billion. The summer of 2017 saw devastating fires ravage Alberta's neighbouring province, British Columbia. In July 2016, the United States Drought Monitor published data showing that over 44 per cent of the contiguous US was experiencing abnormally dry conditions, 16.8 per cent classified as moderate to exceptional drought. The severe drought that gripped the United States in 2012 affected approximately 80 per cent of agricultural lands.

In mid-June 2016, researchers revealed the first mammal to go extinct as a result of anthropogenic climate change. According to scientists, the Bramble Cay melomys, a small rodent endemic to the Great Barrier Reef, was driven to extinction by rising sea levels. A month earlier, a group of Australian researchers published data showing that, as a result of rising sea levels linked to anthropogenic climate change, five uninhabited Pacific islands belonging to the Solomon Islands had been completely submerged, with six more experiencing severe erosion and shoreline recession that in two instances led to the loss of entire villages. In 2014, scientists published separate studies showing that glacial melt on the West Antarctica ice sheet is so significant and rapid that its collapse is largely certain. While possibly taking hundreds of years, the combined studies project 4 metres of sea level rise once the ice disappears, devastating coasts and coastal communities worldwide. More recent research has revealed stunning levels of glacial melt on the Greenland ice sheet, totalling an average of 269 billion tonnes of melt per year between 2011 and 2014, well above annual averages for the previous century.

In June 2016, the National Oceanic and Atmospheric Administration (NOAA) released data indicating that all of the world's coral reefs, from the USA and its territories to the Caribbean, Australia, and beyond, were experiencing the most widespread coral bleaching event in recorded history, occurring when high water temperatures cause the algae that live in the coral to be cast out, turning the coral white. If conditions persist, the coral eventually dies. Research on Australia's famed Great Barrier Reef has shown 'complete ecosystem collapse' in some areas, with around a quarter of the coral now dead, while approximately 93 per cent is experiencing some form of bleaching. Coral reefs globally are estimated to generate $30 billion per year in net benefits from fishing, tourism, food security, and various ecosystem services, including coastal protection during tropical cyclones, which are themselves predicted to become increasingly frequent and more powerful with climate change. In October 2015, Hurricane Patricia, originating off the Mexican coast in the Pacific Ocean, became the strongest hurricane on record with a sustained 1-minute wind speed of 345 kilometres per hour. Patricia took the record from Typhoon Haiyan which struck the Philippines in November 2013, with sustained 1-minute winds of 315 km per hour. Arriving with a storm surge of over 7.5 metres in some places, Haiyan left in its wake over 4 million homeless, more than 6,200 dead, and extensive economic devastation. The highly destructive 2017 Atlantic hurricane season saw three catastrophic hurricanes – Harvey, Irma, and Maria – devastate Houston, Texas, and a string of Caribbean islands. When Hurricane Maria hit Puerto Rico, a massive humanitarian crisis ensued, with the island's energy grid largely destroyed, infrastructure and buildings heavily damaged, basic necessities unavailable, and its agricultural sector devastated. The island nation of Barbuda saw over 90 per cent of its buildings severely damaged or destroyed, the country's government suggesting it could not afford to rebuild. Its citizens were evacuated to the neighbouring island of Antigua, where many now live, given the near-total destruction of the country. It has been estimated that the economic cost to the

United States of wildfires and Hurricanes Harvey, Irma, and Maria, all taking place during the months of August–September 2017, could top $300 billion.[10]

While hardly exhaustive, this sampling of the consequences of a changing climate helps to position climate change in the here and now, as an immediate, tangible, and destructive force, no longer the subject of future projections and future costs. Yet these climate events, or milestones, or catastrophes, are deceptive, for they often appear random and disconnected. A typhoon in the Philippines, wildfires in Canada, rising agricultural unemployment in California – none compels us, necessarily, to connect the dots. Disasters and crises are hardly unusual to the human condition and thus, as one observer highlights, the 'straight lines from cause to effect are rarely if ever visible to the observer. . . . No single decision, no emission of one tonne of greenhouse gases can be connected to this particular scene: the burning of this barrel of Texas oil cannot be pinned down as the cause of this Levantine drought' (p. 4).[11] It might be that we will only really begin to connect the dots once anthropogenic climate change becomes sweepingly catastrophic.

The future of climate change

According to a group of climate scientists who modelled and analysed an exhaustive quantity of data that was then published in the journal *Nature*, the global mean for when 'climate departure from recent variability' is expected to occur is the year 2069 (±18 yrs) under a scenario of emissions stabilization, and 2047 (±14 yrs) under a business-as-usual (BAU) scenario. Expressed differently, *climate departure from recent variability* is the point at which the planet's climate moves outside of the norm, entering territory that is alarmingly 'unprecedented' from the timeframe under analysis (1860–2006); if we quickly stabilize global GHG emissions, which is highly uncertain, we have about fifty-one years from 2018 before we will no longer recognize the climate we live in, but only roughly twenty-nine years if we continue with business as usual. In turn, those numbers represent the mean, or average, of unprecedented climate departures, with the authors highlighting that it will be some of the poorest countries and peoples in the world, concentrated in the equatorial latitudes, that will face the extremes of departure much earlier than those countries and peoples residing in the higher, more temperate latitudes – this as a result of their fairly uniform climates that display minimal variability. Small changes in temperature can have significant consequences because ecosystems and species in these regions have 'adapted to narrow climate bounds', with a much lower tolerance for changes. Since the early emergence of 'historically unprecedented climates' will take place in tropical regions that are host to the majority of the world's biodiversity, the authors underscore the severe threat to this biodiversity posed by climate departure, with high levels of extinction and die-off in both terrestrial and marine ecosystems very likely. The impact this will have on the 1–5 billion people to be affected, depending on the emissions reduction scenario employed, will be equally unprecedented, with the authors stressing the fact that the early emergence of historically unprecedented climates in the world's poorest countries 'further highlights an obvious disparity between those who benefit economically from the processes leading to climate change and those who will have to pay for most of the environmental and social costs'.[12]

In many cases, 'climate divergence' will usher in sweepingly catastrophic conditions that signal a planet changed forever and in ways that are increasingly inhospitable to its inhabitants. Based on the latest scientific projections, we can expect a world of overlapping crises and climate emergencies.[13] Recent research projects that, by the end of this century, temperature increases resulting from anthropogenic climate change in southwest Asia, particularly in the Middle East, will be such that 'certain population centres . . . are likely to experience temperature levels that are intolerable to humans' as a consequence of the physiological inability of the human body to adapt to extreme temperatures beyond a threshold of 35 °C (Pal and Eltahir, p. 197). In other words, the increasing heat in parts of the Middle East is projected to become so intense that it will literally be deadly to the human body. Regions particularly susceptible to extreme heat, in addition to the Middle East, include the Mediterranean, North Africa, the contiguous United States, tropical South America, central Africa, and all tropical islands of the Pacific, amongst others. If global temperatures increase by 4 °C – a real possibility that has inspired a series of reports commissioned by the World Bank to examine the consequences of this level of warming in the regions where it works – 70–80 per cent of the terrestrial land mass in Latin America, the Caribbean, North Africa, and the Middle East, and 55 per cent in areas of Europe and Central Asia, could be affected by 'unprecedented heat extremes'. Recently published research projects that such heat extremes and the resulting incidence of heat stress, heat strain, and heat stroke for vulnerable labourers unable to escape the high temperatures will diminish worker productivity in hard-hit regions, including South-East Asia, by almost 30 per cent by 2050, shaving hundreds of billions of dollars annually from global GDP.

In addition to the consequences to human health and mortality of heat extremes, it is projected that agricultural yields globally in a 4 °C world will be dramatically impacted, with 'significant' effects already well documented at our current level of warming. With a 2 °C increase, it is projected wheat crop yields in Central America, the Caribbean, and Brazil would drop by up to 50%, while soybean yields in Brazil would decline by 30–70%. Studies modelling the impacts of a warming climate and heat extremes on crop yields for corn, soybeans, and cotton in the United States show that, under a scenario of slow warming due to mitigation, yields will decline between 30 and 46%, and by 63–82% under a scenario of rapid warming consistent with business-as-usual fossil fuel combustion. With the United States producing roughly 41% and 38% of global corn and soybean stocks, respectively, a drop in yields, likely to fall somewhere between the two scenarios, poses a severe threat to global food security, particularly as other regions are projected to experience drastic yield declines simultaneously. As the physical scarcity of staple crops increases in line with rising temperatures, global commodity and food prices are expected to rise precipitously, leaving import-dependent nations and poorer and marginalized populations especially vulnerable, and undermining economic growth and development linked to the agricultural sector in many nations. It is estimated that food scarcity resulting from 2–2.5 °C, let alone 4 °C, of warming will lead to 'substantial increases' in malnutrition and stunting in hard-hit regions, robbing future generations of the opportunity for healthy, secure, and prosperous development. In turn, models show that the threats to global agricultural production from soaring temperatures will be compounded by additional climate stressors, including more frequent and intense droughts, the salinization of ground water used for irrigation in coastal areas as sea levels continue to rise, and, in some cases, more extreme precipitation events.

Indeed, dramatic changes to the global hydrological cycle are expected in a warming world, with research showing that the world's wet and dry regions will experience a higher rate of severe rainfall events. This increases the risk of major flooding in impacted regions, with arid regions particularly vulnerable given the absence of infrastructure suited to withstand precipitation extremes. Moreover, severe rainfall and flooding threaten agricultural production, while intensifying the risk and spread of epidemic disease as water supplies are contaminated, thus adding to the burden on critical health infrastructure, particularly in poorer regions. On the opposite side of the hydrological coin, it is projected that global warming will increase the intensity and duration of droughts, drastically reducing water availability in affected regions. Large parts of Africa, North America, South America, southern Europe, and southern Australia are projected to become increasingly dry as global temperatures increase, with estimates showing that 2 °C of warming would reduce mean annual run-off in the Danube, Mississippi, Amazon, and Murray Darling river basins by 20–40 per cent. At 4 °C of warming, it is projected that rainfall levels will decline by 20–50 per cent in the Middle East and North Africa, Central America, the Caribbean, and the Western Balkans, amongst others.

Without a doubt, water turns out to be a central cast member in the unfolding story of anthropogenic climate change. Many of the world's inhabitants will be confronted increasingly by conditions of too much water, or not enough. Extreme precipitation; drought; glacial melt that increases sea levels on the one hand, but diminishes the availability of critical fresh water resources for dependent communities in Andean countries and Central Asia, on the other; sea-level rise from glacial melt and the thermal expansion of ocean water as it heats, inundating coastal settlements and infrastructure globally; ocean acidification and the devastation of marine ecosystems; the list goes on. Many readers of this book may well have seen highly unsettling visual simulations – pictures or videos – of the existential challenge to be experienced by some of the world's most iconic cities as sea levels rise – New York, London, Shanghai. Scientific research has estimated that, with 2 °C of warming, up to 4.7 metres of eventual sea-level rise could be locked in, threatening approximately 130–458 million people globally with displacement, based on 2010 global population figures. At 4 °C of warming, the business-as-usual scenario, sea-level rise of up to 8.9 metres could be locked in, threatening displacement for between 470 and 760 million people globally. Let's think about this – 8.9 metres is equivalent to almost 30 feet, greater than a 2.5-storey building! Countries making the top 20 list with populations and urban centres most vulnerable to sea-level rise include Bangladesh, Brazil, China, Italy, Japan, the UK, the USA, and Vietnam, amongst many more. What these numbers underscore are the particularly severe impacts climate change is having, and will continue to have, on human populations, and that, with little surprise, will contribute to swelling numbers of climate refugees globally.

Climate refugees are made up of those individuals and communities forced to leave, or in many cases flee, their homes, lands, and even nations as a result of climate change and its impacts. In addition to sea-level rise, devastating storms, flooding, drought, and unprecedented changes to local and regional environments and agricultural systems, amongst other problems, are already displacing tens of millions of people annually. When viewed through the lens of the devastating global refugee crisis currently underway as a result of the Syrian conflict and years of war, foreign intervention, and conflict in neighbouring countries, we are given a glimpse into what the future of climate-induced human suffering will look like, particularly given the lack of preparedness – or, in many cases, willingness – of those most capable to shoulder their fair burden of responsibility. As it currently stands, 'climate refugees' or 'environmental refugees' are categories with no legal recognition globally, absent from the UN's Convention and Protocol Relating to the Status of Refugees. While very real and tangible in the air they breathe, the lives they live, and the trauma they experience, climate refugees do not exist in the eyes of international law; states bear no legal obligation to support climate refugees, and such refugees possess no basis for making legal claims. As a result, the victims of climate change, many from the ranks of the world's poorest, and disadvantaged by multiple vectors of inequality – class, gender, race – enter a world of the unseen, to be re-victimized as potential refugee-receiving nations stall on enshrining them as a category of refugee worthy of protection – this for fear of what will come as parts of the planet become increasingly inhospitable.

And, thus, we arrive at the issue of conflict in a warming world. In broaching the topic of the relationship between climate change and conflict, it is important to dispense with some of the more narrow and problematic readings of the link between the two. We hear them in proclamations that tell us the war in the Darfur region of Sudan was the world's first climate change conflict – this from UN Secretary-General Ban Ki-moon in 2007; or in 2015, when Britain's Prince Charles said in an interview:

> Some of us were saying 20 something years ago that if we didn't tackle these issues, you would see ever greater conflict over scarce resources and ever greater difficulties over drought, and the accumulating effect of climate change which means that people have to move. . . . there's

very good evidence that one of the major reasons for this
horror in Syria, funnily enough, was a drought that lasted
for about five or six years, which meant that huge numbers
of people in the end had to leave the land.[14]

The problem with these types of assertions, particularly
prevalent in popular media and certain scholarly and
security circles, is that they tend to erase highly complex
political, economic, social, and environmental histories
that more often than not are marked by deeply unequal
relations of power and wealth, suggesting instead that the
key ingredient fuelling conflict is a changing climate. It
is this erasure that, in turn, eliminates the need to assess
critically what kinds of social relations, political and economic policies and structures, and national and foreign
interests furnish the conditions for conflict to erupt once
the drought hits.

While the drought that hit Syria and neighbouring
nations from 2007 to 2011 was the worst on record, leading
to massive levels of internal migration as agricultural yields
plummeted, the brutal nature of the Syrian regime, its
long history of massive human rights abuses, and its failure to support its citizens adequately in the face of crisis,
played a decisive role in triggering conflict in that nation,
influenced in part by the Arab Spring's pro-democracy
uprisings spreading throughout the region. The failure
to acknowledge these kinds of complexities can serve to
mask the extent to which the interests of powerful elites,
domestic and international, propel conflict, and in many
cases benefit from it. Moreover, it belies the extent to which
such simplifications reduce some human beings, particularly those who are dark-skinned and from so-called foreign
lands, to atomized, self-interested, conflict-prone individuals. Do we imagine the break out of civil war in New York
City or in Norway as the waters rise, or with drought in

California? Or are these narratives largely saved for those we imagine to be, thanks in part to journalists like Robert Kaplan,[15] 'culturally dysfunctional', 'tribal', and 'criminal', living in far-off lands? The responsibility lies with us to reject these simplifications for the harm and violence they do, and because they limit the opportunity for meaningful analyses and dialogue on the impact that climate change has on communities the world over.

So what is the relationship between climate change and conflict? The answer to this question will depend on highly contingent factors from one locale and region to the next. In some cases, the impacts of climate change may induce greater levels of cooperation between communities and nations as they seek solutions to the emergence or deepening of climate-related problems. Transboundary water management is a well-recognized example of resource scarcity prompting high levels of cooperation between states in efforts to manage this scarcity effectively. In other cases, preexisting circumstances, including highly unequal relations of power underpinned by competing claims to resources, may find themselves amplified under the impacts of climate change, which may, in some cases, facilitate an escalation of conflict. What is not in doubt is the fact that climate change will make life exceedingly difficult for more and more people globally, straining national and international resources and services, and destroying life chances for far too many.

The inequality of climate change

Before closing this chapter, it is important – indeed, essential – to reflect on those whose lives will be made increasingly unbearable in the years to come, and whose life chances will be snatched away by the beast of anthropogenic climate change. As the data makes clear, no country or region is, or will be, exempt from the consequences of a warming planet. The United States, Canada, Mexico, Haiti, China, Nigeria, England, Bangladesh – none is escaping the rising seas, warming oceans, destructive rains, and scorching heat. It is this fact that has led many observers to suggest that climate change is the 'great equalizer' – something that affects us all, meaning we all have an interest in acting now to mitigate its worst effects, for our own future and the future of our children. While not entirely without truth, conceptualizing climate change as such elides the obvious fact that some possess much greater wealth, power, and privilege – attributes through which the effects of, and responses to, climate disasters are mediated. When it became apparent that a powerful Hurricane Katrina was going to make landfall in Louisiana in 2005, with New Orleans suffering a direct hit, those disproportionately devastated by the storm were African American, poor, unemployed, and living in rental accommodations. When orders came from the government to evacuate, it was these citizens who were least able to comply, given their constrained resources. As well, it is citizens within these demographic categories who are least likely to be covered by insurance, necessary for beginning to rebuild once the storm has passed.

During the summer of 2016, countries in southern Africa were suffering under a crippling regional drought, brought on by the most severe El Niño event on record. The severity, linked to climate change, led to a full-blown humanitarian crisis as crops failed, livestock died, water sources dried up, and widespread food shortages ensued. Estimates suggested that at least 36 million were in need of immediate humanitarian assistance, with an additional 13 million suffering food insecurity at the time of writing.

Whether it's New Orleans, Zimbabwe, Mozambique, or Malawi, this is a scenario that is playing out with increasing frequency the world over as climate-related disasters hit. We need not look too far into any particular event to see what they all have in common, to varying degrees – poverty, insecurity, racial and gender inequality, and so forth. Climate change acts to amplify and deepen preexisting inequalities, given that each reduces the capacity of the individual and the community to withstand and respond to climate shocks. It is thus that the inequality of climate change is both irony and tragedy, for it is those least responsible for anthropogenic climate change that are paying, and will continue to pay, the steepest price. Their voices, while not without strength, nevertheless confront profound structural barriers to being heard and taken seriously, particularly within domestic and international political systems that remain dominated by powerful elites, many of which have a stake in maintaining business as usual. No, climate change is not the great equalizer. Rather, its impacts filter through the complex and often highly unequal social, political, and economic structures that define human social systems globally. And not surprisingly, efforts to mitigate climate change and ameliorate its impacts are likewise channelled through those very same structures; indeed, climate change has emerged out of them. It is to this history and to the global political economy of carbon that we now turn, so that the reader, familiar with the science of climate change, and its threats and impacts, might better understand why translating science into meaningful action of the kind able to avert catastrophic warming has proven so utterly difficult.

CHAPTER TWO

The Global Political Economy of Carbon

Carbon as CO_2 is little more than a molecule, forged with the bonding of oxygen and carbon atoms. The basic problem of anthropogenic climate change is one of numbers – there are now too many molecules in our atmosphere, requiring serious work to figure out how we might dramatically limit adding any more. And yet the problem of too much carbon emerges out of specific political economic and socio-cultural dynamics. If carbon is reduced to its molecular base as a point of departure for understanding the problem, these dynamics are less clearly visible, while carbon's capacity to transform much more than the natural world may be altogether obscured. As this chapter was being prepared, all evidence indicated that 2017 would be a year of climate milestones: likely the warmest non-El-Niño year on record and the year with the highest-ever human-driven greenhouse gas emissions. Both milestones have since been confirmed. Researchers suggest that emissions must peak by 2020, beginning a rapid decline thereafter, while evidence now exists to suggest that the most accurate climate change models are also those that anticipate the most dangerous levels of warming.[1] In all of this, we glimpse what it is that makes climate change different from so many other problems, and that is the issue of time. Indeed, climate change bears a particular temporality whose urgency requires rapid and deep change; and while change is happening, it is hardly commensurate with the scale of the challenge that lies ahead. To understand why, we must look to the political economic and socio-cultural dynamics within which too much carbon became possible. This requires that we appreciate the central role played by fossil fuels in powering industrial growth and development, beginning in the eighteenth century. More than that, it requires an appreciation of our current economic system and its basic imperatives. As a system of production, capitalism requires continuous growth and profitability. While fossil fuels have been uniquely suited to meeting these imperatives, a system that itself requires limitless growth on a limited planet was bound to run headlong into trouble. Although hardly visible throughout much of this history, carbon has been ever present, powering lives and generating very specific political and cultural formations that make it spectacularly difficult to kick the carbon habit. We must thus examine this history closely as a starting point for understanding how we got here. The better this task is fulfilled, the better prepared we are to think about where we need to go.

To grow or die

To speak of economic decline – crisis, recession, depression – is to conjure up the spectre of death. On a cognitive level, economic decline smashes illusions of progress, security, and vitality since, in the realm of the material, economic decline strips its victims of that which is required for life under a capitalist system. Economic decline heralds job losses, unpaid bills, burgeoning debt, foreclosed mortgages, medicines forgone, empty stomachs. Our dignity and our humanity depend deeply on the fortunes of the economy – does one have a job? Is it a good one? Does it pay enough to cover daily expenses? Does it come with benefits and job security? In the best of economic times, the number of citizens answering 'no' to these kinds of questions is unacceptably high, but when times get tough and economies contract, their numbers balloon, with far too many pushed over the edge. It is in this way that the human body and its capacity for life are bound to the rhythms and requirements of the capitalist economy, and thus it is here that we will begin our discussion of the global political economy of carbon.

What distinguishes the capitalist market economy from other forms of economic organization is its driving imperative. Unlike markets for bartering or those that are socialist in orientation, capitalist markets are organized around the realization of profit – the difference in value between the cost of producing a good or service and its sale price. If the car manufacturer were to charge less for the vehicle than it cost to make it, or simply to break even, or if the smartphone company were to sell its devices at a loss or at cost, ruin for each would quickly follow. Businesses in a capitalist market economy are not motivated by one-off opportunities, never to be repeated. Rather, a capitalist economy is always moving; it requires investment for the purpose of profit realization, summed up as making more money than the initial amount invested. As a system in motion, the point is to reinvest a portion of those profits back into the same process – making cars, for example – in order to reproduce the dynamic on ever-larger scales. What would happen to the CEO who announced company plans to minimize growth in the coming fiscal year, as a precursor to the company's future zero-growth strategy? Indeed, any such strategy would mark the company's exit from the capitalist system of production, for capitalism is a system of profit maximization through growth. Moreover, profit maximization requires minimizing the costs of production

– the wages and benefits paid to labourers, the amount paid to operate the factory, and the cost of the inputs from which a product is produced. With competitors waiting in the wings to steal market shares – to which the woes of Blackberry, Polaroid, and IBM can attest – a company's survival depends on its ability to out-compete the competitor. For those unable to do so, the punishment is swift – markets dry up, share value plummets, workers are cast out.

In this way, the *grow or die* imperative specific to the capitalist system of production has dual significance. Most obviously, if the company fails to grow according to the rules of the game, it dies. Perhaps less obvious, depending on one's vantage point, is the spectre of death alluded to above. The worker cast out in a fully commodified society is confronted with the prospect of an untimely end if a new source of money cannot be found. In a capitalist society, one's bodily survival is contingent on access to money as the medium of exchange through which the necessities of life – food, clothing, shelter, and healthcare – are secured. The level of commodification in a capitalist society, or the degree to which all objects, including those necessary for social reproduction, are 'produced for sale on the market' (p. 72),[2] will determine the distance between life and death. Where healthcare is predominantly commodified and for profit, as in the United States, more citizens die from being unable to pay for life-saving treatments or medications – with the poor and minority communities heavily over-represented – than in nations where healthcare is treated as a public good. Statistics show that, in both rich and poor nations, life expectancy at birth is much lower for those unable to pay for the necessities of life. Again in the United States, this fact is borne out in recent data showing that poorer women born in 1940 and residing in the bottom 10 per cent of income earners die on average ten years earlier than rich women in the top 10 per cent; that gap widens to twelve years for men.[3]

While this discussion may be perceived as a detour from the goals of detailing carbon's complex political economy, the story of carbon is one that cannot be told without first unpacking the socio-economic imperatives that shape our current economic system. Richard Peet, Paul Robbins, and Michael Watts, paraphrasing the writing of Antonio Gramsci, note that people across the globe today, the majority of whom do not own companies but rather must secure work as labourers within them, find their lives dominated by the 'real necessity of production' (p. 15).[4] The need to find paid work and to get a job is a 'brute fact' if one is to survive, let alone thrive, in a capitalist society, since acquiring what we need requires money. This relates in important ways to how commodification is itself underpinned by the enclosure and privatization of land and resources, and of goods and services, as a basic requirement for capitalist ownership. The consequence of transforming common or public resources into private property and assets to be sold for profit is that individuals are, quite literally, separated from those resources and goods that they need to live. In his book *Fossil Capital: The Rise of Steam Power and the Roots of Global Warming*, Andreas Malm opens chapter 13 with a particularly useful analogy. He states:

> Woodpeckers work on the excavation of wood. The bills are their tools. Striking their sharp-nosed hammers with a signature mechanical sound, they can bore mouth-sized holes into tree trunks, like shafts for mining ants and termites, beetles and their grubs. Because the tools are at one with the bodies of the birds, they cannot be concentrated. No master woodpecker can collect bills and pile them up on a central site and tell the other members of the population,

their faces strangely flat, to submit to his command and get access to the tools they need to break through the bark or refuse and starve in freedom: for this reason, if for no other, property relations among woodpeckers are impossible. Their equipment for metabolism cannot be distributed between owners and non-owners, nor can it be collectively controlled by a commune.[5] (p. 279)

In other words, there is no separating woodpeckers from that which they need to survive – they own their bills and so they eat. Not so for humans under capitalist social relations. For so many the world over, someone else, driven to profit, owns what we need to survive, and in order to be able to access it, we must work to make money, since that is the only way it is coming to us.

Viewed this way, the difficulty of disentangling the needs of workers from those of capital becomes clear. If capital doesn't survive and the economy doesn't grow, then our bodily survival as human beings who must work to live is increasingly uncertain. Of course, there is a wide gulf between the interests of capital in general and those of workers since capital is driven by the imperative to maximize profitability, often pursuing cost-cutting strategies that diminish the life opportunities of workers. Nevertheless, workers, like capital, need growth, meaning that workers are largely ensnared in, and dependent upon, a system whose basic growth imperative is undermining many of the planet's critical life-support systems, including its climate.

Consider that the Italian auto company Lamborghini announced in late 2017 that it was introducing a luxury SUV, moving into mostly uncharted territory for the company famed for its luxury sports cars. While the starting price of US$200,000 ensures this will remain a commodity for the elite, it is the market trends that inspired the company's decision that matter to the story of carbon, capitalism, and growth. Whether looking at Lamborghini, Porsche, General Motors, or Ford, observers note that the continued expansion of the market for trucks and SUVs globally, the most fuel-inefficient of all passenger vehicles, compels auto manufacturers to produce these vehicles at exceptional rates, the majority still powered by the carbon-intensive internal combustion engine. Statistics show, moreover, that between 1990 and 1999, 32.2 million new cars were sold worldwide and, between 2000 and 2013, that number inched up to 53.73 million. Reflecting massive growth in emerging markets, 2014 alone saw 71.18 million cars sold globally, jumping to 77.31 million in 2016. It is projected that 2017 will see a new high of 78.59 million.

While the growth of the electric vehicle market globally garners significant attention for the contribution it might make to lowering global GHG emissions, a recent report from the International Energy Agency (IEA) noted that, in 2016, electric vehicles accounted for only 0.2 per cent of all passenger light duty vehicle sales globally. Responding to the data, the report's authors noted that 'electric vehicles still have a long way to go before reaching deployment scales capable of making a significant dent in the development of global oil demand and greenhouse gas emissions'. The global transportation sector, which also includes air travel, is responsible for 14 per cent of all GHG emissions worldwide; that figure is around 20 per cent and 24 per cent in the USA and Canada, respectively. With each year's new vehicle models, with leasing schemes that have consumers trading in the old for new every five years, and with strategic efforts to capture the fastest-growing markets in emerging economies – perceived as massively untapped reservoirs of potential new drivers globally – the auto industry is banking on a future of growth to sustain it. Employing many millions of workers globally, this is also a strategically important sector for the jobs it provides and the bodies it sustains. Finally, the global auto sector provides important insights into issues of governance, policy, and the cultural politics of carbon. Around the world, governments actively support their auto sectors with favourable policies and the development of infrastructure conducive to encouraging more driving – we see this in subsidies and tax breaks to the sector, in how cities and suburbs have been constructed historically in North America, and in the underfunding of efficient public transportation networks. Additionally, however, vehicle ownership confirms particular narratives – status, freedom, convenience, modernity. In all of this, the dense web of personal automobility helps to frame our understanding of why dealing with too much carbon is so incredibly daunting.[6]

From the early development of capitalist modes of production, those who owned the factories and the growing businesses were confronted with the necessity of achieving self-sustaining growth and profit maximization. Since neither of these emerge naturally, identifying those factors and conditions that either promote or impede their realization was and continues to be a central task of the capitalist firm. Are there limits to the amount of goods that can be produced in a given period of time? Are the necessary factors of production accessible and available? Do inputs cost too much? What impact do existing policies have on operations and production? While the answers to such questions vary through time and space, fossil fuels have been and continue to be critical to securing growth and profitability, meaning that carbon, as a by-product of fossil fuel combustion, is a central cast member in this drama. So what is it about fossil fuels?

Perhaps most obviously, production requires energy.

How do the tractors move about their fields and harvest their bounty? Through what process can heat sufficient to produce cement and steel be generated? What powers the drills that build our cars, the sewing machines that produce our clothes, the saws that cut the trees, or the boats that catch the fish? In 2014, 81.1 per cent of the world's total energy supply was made up of fossil fuels. In its 2016 *World Energy Outlook*, the International Energy Agency modelled continued growth in fossil fuel use through to 2040, estimating that global energy demand would increase by 30 per cent over that period. With this in mind, the report's authors offered this troubling reflection: 'For the moment, the collective signal sent by governments in their [Paris] climate pledges . . . is that fossil fuels, in particular natural gas and oil, will continue to be a bedrock of the global energy system for decades to come' (p. 5).[7]

Capitalism, fossil fuels, and carbon
Indeed, fossil fuels possess specific characteristics that are aptly suited to answering the call of growth and profit maximization, and that, in some cases, have less to do with their energetic properties than with the social relations they make possible. As Malm highlights in *Fossil Capital*,[8] historical accounts of the adoption of steam (coal-fired) power in early industrializing Britain tend to suggest that resource scarcity was decisive in fuelling the emergence of fossil-based power. Proponents of this perspective suggest that, had British producers chosen to rely on trees to provide wood that could light the fires necessary for industrial production, self-sustaining growth would not have been possible as, fairly quickly, every last tree in Britain would have been cut. According to the same logic, had producers continued to rely on water to turn the wheels that powered their cotton and textile mills, no inch of suitable space

would have been left on the river banks upon which to build even one more mill. Of course, these are hypothetical scenarios that never came to pass; the water that powered the wheel at the heart of Britain's emerging cotton industry in the latter years of the eighteenth century was both plentiful and cheaper than the coal that displaced it. The appeal of coal as a source of fuel to power the steam engine had much more to do with the power it afforded capitalists to control labour as a critical input into the production process; an *input* whose actions could significantly constrain growth and profitability.

First, the steam engine allowed for the automation of production, replacing human workers with machines. When in the 1820s and 1830s, spinners in the cotton industry began to organize and push wages higher despite a deep structural crisis in the sector, automation allowed mill owners to shed the workers. When hand-loom weavers, of whom there were many, found themselves working up to 18 hours a day in the 1830s to make up for the dramatic cut in their wages, they turned to stealing yarn for a burgeoning black market in order to survive. Again, automation allowed factory owners to rid themselves of their reliance on human labourers and the contingency, uncertainty, and reduced profitability that they brought with them. Automation also allowed for increases to productivity, and thus growth. Beyond that, steam power possessed distinct benefits in space and time. Relying on water to power cotton mills meant that owners had to go to where the water flowed. Often far from urban centres where labour was abundant, owners were especially vulnerable to labour shortages and unrest, while at the same time investing significant capital to attract and hold on to the workers they did have. In fact, owners began to rely heavily on forced labour, including that provided by impoverished children who had wound up in poor-houses, to fill their ever-growing need. When the Factory Movement of the 1830s began to gain ground in its battles for a shorter working week, the superiority of steam power relative to water was further underscored. Water could be erratic in supply, depending on variables such as weather and the time of year, requiring considerable flexibility for mill owners to operate when conditions were optimal. Once the government began to legislate and enforce rules on child labour, working conditions, and shorter workdays, that flexibility was lost. Using coal to produce steam, on the other hand, could be done anywhere, anytime, effectively freeing capital to expand on its own terms. And thus the fossil economy was born, not as a result of resource scarcity, but given its superior ability to facilitate capital's basic imperatives of growth and profitability – imperatives threatened by democratic mobilization on the part of an emerging labour movement.

In this respect, the story of oil is strikingly similar to that of coal. As coal became the dominant source of energy for the British, European, and American industrial machines in the late nineteenth and early twentieth centuries, the unacceptable working conditions of coal miners and those along its supply chain – railway workers, port workers, and the like – fuelled worker mobilizations to make demands for improvements in the conditions under which their labour was carried out, and for democratic reforms. Given the strategic importance of coal to the production process, labour unrest threatened not only coal companies, but companies throughout the economy whose production depended upon its steady supply. As Timothy Mitchell notes in his book *Carbon Democracy: Political Power in the Age of Oil* (2013),[9] the sites where abundant quantities of high-grade coal were found in the late 1800s and early 1900s were relatively few – select areas in France, England, Wales,

Belgium, and in Appalachia in the United States. In order to get the coal from mine to market, and from market to end user, 'narrow, purpose-built channels' – railways, bridges, ports – developed, along which workers were able, 'at certain moments, to forge a new kind of political power' (p. 19). Indeed, Mitchell highlights that, unlike other workers, those operating along the coal supply chain possessed extraordinary political power by virtue of the sheer 'quantities of carbon energy' that they controlled, and that 'could be used to assemble political agency . . . to slow, disrupt or cut off its supply' (p. 19). With reference to the proliferaton of strikes in the mining sector during this period, he continues:

> The flow and concentration of energy made it possible to connect the demands of miners to those of others, and to give their arguments a technical force that could not easily be ignored. Strikes became effective, not because of mining's isolation, but on the contrary because of the flows of carbon that connected chambers beneath the ground to every factory, office, home or means of transportation that depended on steam or electric power. (p. 21)

In short, without this vital source of energy, the capitalist economy would grind to a halt. And it is here, at the intersection of energy, profit, and labour rights, that the appeal of oil as a new source of productive energy began to grow as the twentieth century wore on. Compared to coal, the oil supply chain requires fewer workers – no underground mining is required, oil can be transported in pipes, or in ocean tankers that themselves run on oil with no need for the coal 'heavers and stokers' that worked the trains. Moreover, transporting oil across the oceans allowed companies to operate beyond the reach of national labour laws and democratic gains that had been achieved in relation to coal. When the US government launched the Marshall Plan for European post-war reconstruction following the devastation of World War II, a significant portion of its funds financed the purchase of oil produced by American companies. While strategically beneficial to US economic interests, Mitchell notes that one of the principal goals of this approach was to break the 'political power of Europe's coal miners'. Between 1948 and 1960, oil went from meeting 10 per cent of European energy needs to meeting nearly one-third. As the Middle East rose as a strategic source of oil globally throughout the twentieth century, oil companies were able to exploit these advantages further, fostering highly undemocratic forms of governance, and helping to cement the dominance of fossil-based and carbon-intensive energy as the primary input into the capitalist system of production. Indeed, for anyone interested in understanding the history of fossil fuels as a source of productive energy driving capitalist economic growth, it makes sense to start with the issue of labour as a critical input into an economic system that demands self-sustained growth and profitability, but one that, under the right conditions, has the power to limit or block their realization.

Of course, for a system built on the imperatives of limitless growth, the superior value of fossil fuels for achieving this end would become increasingly apparent over the course of the industrial revolution, and throughout the nineteenth and twentieth centuries as capitalism matured. In particular, fossil fuels possess a very distinct materiality marked by a remarkably high density of energy that has allowed for a high Energy Return on Energy Invested (EROEI). EROEI refers to the amount of energy that must be invested in order to produce usable energy – the fires, steam, and heat capable of powering the factories, driving the cars, and keeping the lights on. With the rise of fossil-fuel-driven production, it became increasingly clear that coal, oil, and natural gas yielded significant amounts of energy with fairly minimal exertion in the process of extraction. The quantities of energy made available from fossil fuel combustion, starting with coal, rivalled those of renewable sources with the onset of the industrial revolution, providing fossil-based fuels with a distinct advantage over their organic counterparts (p. 42).[10] When combined with the fact that fossil energy freed producers from the shackles of time and space discussed above, allowing the energy derived from these carbon-intensive fuels to be harnessed anytime, anywhere, we can see clearly how it is that fossil fuels emerged as superior sources of energy to power capitalist production. And while trends in the renewable energy sector are encouraging, carbon-intensive fossil fuels continue to dominate the global energy landscape. Of course, one of the profound contradictions of our dependence on fossil fuels is that, as sources of highly potent energy, they are remarkably inefficient. While solar and wind energy can be harnessed for productive use immediately, something that technological advancements are making easier by the day, fossil fuels as 'buried sunshine' took hundreds of millions of years to form deep underground and, once used, they're gone. These types of inefficiencies matter little, however, in the pursuit of self-sustained growth and profitability. At the end of the day, getting the most bang for your buck, so to speak, has been the name of the game.

The age of growth

An overview of global economic growth, measured in gross domestic product (GDP), shows that, up until the first decades of the nineteenth century, it remained consistently low. It is around 1820 that we begin to see a fairly 'spectacular' change in global growth rates; average per capita GDP in the 1820s stood at around $650, rising to nearly $1,300 in the first years of the twentieth century, then on to $2,405 in the 1950s, and reaching nearly $7,000 in the early 2000s. This represents a tenfold increase between 1820 and 2010. What changed in the 1820s was that capitalist industrial relations, as a result of the industrial revolution, had begun to take root sufficiently. As a result, the logic of growth and profitability gained prominence in organizing economic decision-making and social relations, particularly in Britain and then expanding outward, especially to other Western European nations and the United States. What must be underscored in any account of global economic growth, however, is how incredibly uneven and unequal growth has been, throughout history and as a result of the economic system driving that growth. This point is illustrated in table 2.1 which lists GDP per capita figures for select countries, beginning in 1820. While the GDP per capita in the USA in 2010 stood at over $30,000, the World Bank estimates that the average GDP per capita in the world's poorest countries stood at a deplorable $617 in 2015.

There is a tendency amongst some to suggest that it has been a mix of superior Western characteristics and traits that led to development and modernization in the West, meaning that the absence of these same characteristics is responsible for underdevelopment and poverty in much of the global South. While paternalistic, ethnocentric, and

Table 2.1 Per capita global GDP (US$)[11]						
	Britain	United States	Japan	China	India	Nigeria	Global
1820	$2,074	$1,361	—	$600	—	—	$605
1900	$4,492	$4,091	$1,180	$545	$599	—	$1,225
1950	$6,939	$9,561	$1,921	$448	$619	$753	$2,082
2010	$23,777	$30,491	$21,935	$8,032	$3,372	$1,876	$7,890

often racist in their assumptions, discourses such as this simultaneously sanitize the history of capitalist social relations globally, such that we fail to see how the drive for limitless growth and profitability led to a specific historical geography of capitalism marked by exploitation and subjugation on a global scale. Once carbon-based energy was well established, the earliest industrializers were soon confronted by additional limits: rapid industrial expansion demanded more and more resources to fuel production; the growth of cities as spaces where industry concentrated meant fewer were engaged in agricultural production to feed hungry urban bodies; and limited consumer capacity led to the saturation of domestic markets. For many reasons, these industrial centres required what other regions of the world had. Colonizing territories and peoples around the world; privatizing, extracting, and stealing their resources; turning communities and families into slaves to be brutalized and sold like common cattle; and demanding servitude so that imperialist nations could prosper – these were the prices to be paid to feed limitless expansion. Accompanied by discourses of European and American cultural and racial superiority, which lent credence to the belief that colonialism offered the 'gift' of civilization and modernity to the world's 'backwards' and 'barbaric' peoples, it is the historical processes of colonialism and imperialism, and the specific geography that came with them, that set in motion the evolution of capitalist economic growth as a highly uneven system dependent on exploitation. Moreover, colonial domination ensured that the colonies themselves did not emerge as competitors. If given the freedom to develop their own carbon-intensive industries in need of resources, workers, and markets, economic development in imperial nations would most probably have succumbed to the limits discussed above.[12]

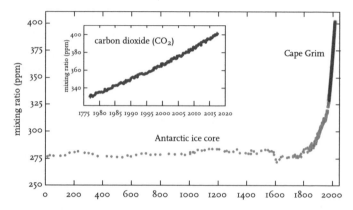

Figure 2.1 Concentrations of CO_2 in the atmosphere (ppm) –
Years 0–2017

Source: Paul Krummel and David Etheridge, CSIRO Oceans and Atmosphere
– Climate Science Centre, 2017.

When the highly uneven historical geography of global
capitalist economic growth is illuminated, a spotlight is
shone simultaneously on the highly uneven historical
geography of carbon, given the bonds that connect the two.
Just as economic growth in today's rich nations took off in
the 1800s, so too did their emissions of carbon dioxide as
a by-product of fossil-fuel-dependent development. Global
data shows that it was during the 1800s that the concentration of CO_2 in the atmosphere began its dramatic ascent,
from about 285 ppm in 1850 to over 400 ppm today, break-
ing with at least 800,000 years of relative stability prior to
the onset of the industrial revolution (see figure 2.1).

As the world's nations began to grapple with the threat
of catastrophic climate change and the recognition that
combatting it would require a dramatic reduction in CO_2
emissions, issues of equity, responsibility, and fairness soon
dominated debates. Indeed, many in the global South asked

why they should be held to the same standard as their rich counterparts in the global North when it was the world's rich countries that were largely responsible for some 200 years of carbon emissions, while many poor and emerging economies were still struggling to overcome the legacies of colonialism and achieve rates of growth and development – albeit carbon-intensive growth and development – comparable to those in the West. As we will see below, it is this question of *historic responsibility* that became a focal point around which an emerging global regime for climate governance would, in many ways, coalesce.

Climate debt and historical responsibility
In 1992, leaders from around the world met in Rio de Janeiro, Brazil, for the United Nations Conference on Environment and Development (UNCED), dubbed the Rio Earth Summit. Unprecedented in size and the number of issues that it sought to address, with 172 governments and 108 world leaders in attendance, and thousands more from civil society and the private sector, it was at the Earth Summit that the UNFCCC was born, ushering in the first global Convention designed specifically to facilitate international action on climate change; today, the UNFCCC boasts a membership of 197 countries. It was at this historic 1992 meeting that George Bush Sr., then President of the United States, famously asserted: 'The American way of life is not up for negotiations. Period.' Reflecting the highly contentious nature of the meeting that often saw rich nations pitted against their poorer counterparts, Bush was both aggressive and arrogant in making clear that, while action may be required on global environmental issues, it would not come at the expense of the mass consumer lifestyles that had come to define the United States, even if enjoyed in highly uneven ways as a result of class, gendered, and racial

divides in that nation. It would be under Bush Sr.'s son, President George W. Bush, that the United States would refuse to ratify the Kyoto Protocol, the first global treaty of its kind to set out specific emissions reduction targets and timelines for achieving them. After President Barack Obama's two terms in office (2009–17), during which he signalled a willingness on the part of the USA to engage in global efforts to combat climate change, a President hostile to the very idea of anthropogenic climate change now sits in the White House. While it is easy enough to trace this hostility back to the main obstacle impeding effective action on climate change – in this case, the centrality of fossil fuels to American economic growth and power – it simultaneously shines a spotlight on the intractability of adequately accounting for historic responsibility.

The Climate Equity Reference Project, co-managed by the think tank EcoEquity and the Stockholm Environment Institute, has been designed precisely to wade into the stormy waters of historic responsibility. Drawing on climate science from the IPCC, alongside historic data on economic growth and development globally, the Project offers tools for analysing what an equity-based approach to climate action would look like if countries were to take on their fair share of responsibility. Their Climate Equity Reference Calculator allows users to select the parameters for assessing fair share, including mitigation required for a 2 °C or 1.5 °C pathway, the base year from which emissions should be measured, and how much weight historic responsibility versus capability to mitigate should be given. Based on settings that target a 2 °C mitigation pathway in accordance with the commitments laid out in the Paris Agreement, with a base year of 1850 to coincide closely with the onset of rapid industrial growth in the West, and a weighting of 100 per cent responsibility over capability

Table 2.2 Calculating historic responsibility to stay below 2 °C of warming		
Region/country	Allowable emissions above/below projected 2030 levels (%)	Allowable emissions above/below 1990 levels (%)
World	−43	+13
Western Europe	−131	−130
United Kingdom	−157	−146
United States	−164	−178
China	−11	+424
India	−0.61	+339
Brazil	−60	−39
South Africa	−25	+43
Least Developed Countries	−2.0	+76

Source: Climate Equity Reference Calculator, available at https://calculator. climateequityreference.org.

to reflect who has emitted the most carbon into the atmosphere historically, we can calculate who bears the greatest responsibility for current levels of warming, globally. The results, provided in table 2.2, illustrate for selected countries and regions the emissions reductions – or allowances, in a few cases – that would be required to stay below 2 °C of warming above pre-industrial levels, using the years 2030 and 1990 as baselines.

These figures tell us a number of things. In 1990, there was still room to increase emissions globally and meet a 2 °C target, although, if determined equitably, the space to grow would have belonged to countries in the global South. This is no longer the case for 2030, based on current emissions projections. If the United States was to take on its full historic responsibility in combatting climate change, it is estimated that it would have to lower its emissions

164 per cent. This would require lowering its emissions to zero, while supporting reductions elsewhere worth 64 per cent of its own emissions. Countries like China and India, given their relatively recent entrance into the high-emitters club, bear much less responsibility. These numbers contrast quite starkly with how responsibility is often framed in popular discourse in North America. Where I live in Canada, it is fairly common to hear citizens reference our size – a mere 36 million people – relative to that of China at 1.4 billion, or India at 1.3 billion, to suggest that our contribution to climate change is negligible and that it would be unfair to expect us to lower our emissions when the real focus should be on the 'mega-polluters' in the Third World.

Arguments of this nature also fail to account for the issue of emissions embedded in trade. At the level of global governance and the UNFCCC, data is collected based on national emissions inventories – i.e., how much is being emitted within each individual nation. However, when we consider the carbon emissions embedded in the imported goods a nation consumes rather than those resulting from domestic production, we see that many countries in the global North have higher carbon emissions than are otherwise reported. This phenomenon has become more pronounced during the era of neoliberal globalization, with recent research suggesting that approximately 22% of global carbon emissions are the result of goods produced in one country but consumed in another. Tellingly, data shows that, while the UK reported emissions reductions of 27% between 1990 and 2014, once imported emissions were accounted for, that nation's emissions reductions were cut to only 11%. The same adjustment shows that the USA's reported 9% increase in emissions between 1990 and 2014 goes up to 17% once emissions embedded in trade are accounted for. In 2014, China's national emissions were 13% lower once exported CO_2 emissions were taken into account. This paints a more complicated picture in terms of determining both historic and present responsibility for emission.[13]

In reality, the collective data points to a past, present, and future within which the world's wealthy nations and citizens have claimed, and continue to claim, a disproportionate share of the world's available carbon space – that limited space within the planet's atmosphere into which carbon can be emitted without sending us off the climate change cliff.[14] Of course, some have already fallen off the cliff, while others may be able to avoid it indefinitely, given the wealth at their disposal. If we take the politically agreed-upon 2 °C limit as our yardstick, climate science from the IPCC and leading global experts estimated in 2016 that a total of 2,900 gigatonnes of CO_2 ($GtCO_2$) could be emitted if we are to have a >66 per cent chance of staying below 2 °C. Taking 1870 as their start year, their data showed that, through to 2016, approximately 2,100 $GtCO_2$ had already been emitted, leaving a mere 800 $GtCO_2$ to go. This carbon budget – 800 $GtCO_2$ – represented the world's remaining carbon space, with estimates suggesting that, at current levels of emissions, that space will disappear within a mere twenty years (see figure 2.2). If we take 1.5 °C as the limit past which warming cannot go – this based on the already devastating climate change impacts that are being experienced on an annual basis globally and which were discussed in chapter 1 – then that available carbon space shrinks even further.

Based on the historical record, a fairly compelling case emerges that the world's remaining carbon space belongs to those nations and peoples who have yet to claim their fair share, and have reaped few of the benefits that have come from carbon-intensive growth and development. Even if

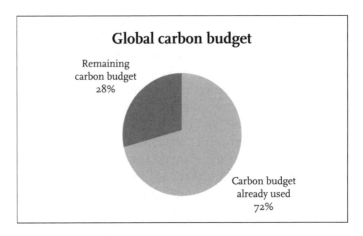

Figure 2.2 Remaining global carbon budget to stay below 2 °C

Source: adapted from the *Global Carbon Budget 2016*, available at www.
globalcarbonproject.org/carbonbudget/16/files/GCP_CarbonBudget_2016.
pdf.

they were successful in claiming this remaining carbon space, however – a scenario that remains entirely out of reach, given the political economy of action on climate change and the North's continued dependence on fossil fuels to power every facet of life – the climate debt owed to the global South would remain. Indeed, if we imagine the last 200 years as one giant party, we know that it has been the world's advanced industrial nations and the wealthiest members of society that did most of the binge drinking, raiding the liquor cabinets of their poorer counterparts and trashing the place, only now telling them that the party's over and they can't replenish the stocks. This underscores a set of highly uncomfortable facts.

In order to avoid truly catastrophic climate change, the carbon-intensive development trajectories of the world's rich nations and peoples cannot be reproduced, yet the historical record shows us that no country has experienced rapid economic growth without a similarly dramatic increase in emissions. The United Nations Development Programme (UNDP) titled its 2013 Human Development Report *The Rise of the South* to draw attention to the unprecedented levels of rapid economic growth underway in some nations in the global South, often grouped as the BRICS (Brazil, Russia, India, China, and South Africa) or BASIC (Brazil, South Africa, India, and China) countries, amongst others. The growth in carbon emissions in these so-called emerging economies has been significant: between 2000 and 2011, China and India together accounted for 83% of the global increase in GHG emissions. At the national level between 1997 and 2008, emissions grew by 103% in China, 67% in India, 40% in Turkey, 32% in Mexico, and 17% in South Africa. The BASIC countries, alongside other key emerging economies including Mexico and South Korea, now control over one-third of global emissions.[15]

While, certainly, renewable energy sources are more readily available and cost-effective today than just a few short years ago, fossil fuels persist as our main source of energy globally. Recent data shows that, while coal is making up a smaller share of overall global power generation, the 84 gigawatts (GW) that were added to global capacity in 2015 represented a 25 per cent increase over 2014 levels. Between 2010 and 2015, the world saw 473 GW of coal capacity added to the sector, with over 90 per cent of the new coal-fired power plant builds occurring in Asia, with China and India leading the way. If left there, with not one more coal-fired power plant built anywhere in the world, ever, the emissions from existing capacity would have been 150 per cent above what the science tells us is needed to keep warming below 2 °C. Of course, we didn't stop there, with significant new capacity coming online in

2016, largely in regions and countries of the global South.[16] And none of this accounts for existing and new capacity in oil and natural gas, bringing us to another uncomfortable fact: while the construction of new coal capacity is on the decline in countries of the global North, the picture is quite different for coal's other fossil-based and carbon-intensive siblings. As natural gas infrastructure expands rapidly throughout the advanced industrialized world, Donald Trump has issued executive orders to green-light the construction and completion of two highly controversial oil pipelines – Keystone XL and the Dakota Access pipeline. Keystone XL, deeply contested because it will transport oil from Alberta's highly carbon-intensive tar sands – which, according to extensive modelling, must remain underground if we are to have any chance of staying below 2 °C of warming – was cheered on by the government of Justin Trudeau in Canada, with the Trudeau Liberals themselves approving additional controversial pipelines in an effort to get Canada's fossil fuels to export markets in Asia and elsewhere. So, in fact, the party's not over, with some of the worst offenders in terms of historic responsibility determined to keep it going.

Before turning to examine global climate governance, it is worth briefly considering the phenomenon of 'decoupling', whereby economic growth is decoupled from the growth in carbon emissions. In other words, they do not grow in tandem alongside each other. Global data points to mostly flat carbon emissions between 2014 and 2016, with global economic growth of around 3 per cent during that time. Observers have also pointed to the United States to highlight similar trends. This data has led many to suggest this is a major transformation, since we are seeing signs that growth need not come with increased carbon emissions. However, preliminary data for 2017 shows that global carbon emissions are rising once again, by about 2 per cent for 2017, which is quite significant. Moreover, observers urge caution in examining the US experience, noting that one of the reasons that country's limited economic growth between 2008 and 2015 did not entail increased carbon emissions was that the carbon intensity of the electricity sector declined as natural gas replaced coal. While we are seeing important changes to the energy sector globally with the expansion of renewables, the resumption in the growth of global carbon emissions and the continued dominance of fossil fuels to meet global energy needs requires that we exercise significant caution when considering the phenomenon of decoupling. As one researcher has noted, while some decoupling may indeed be happening and will probably continue, the data suggests that the decoupling will remain insufficient to prevent dangerous global temperature increases.[17]

The politics of climate governance

This chapter's historical account of the global political economy of carbon deals largely with what Bridge refers to as the 'old carbon economy', rooted in fossil fuel extraction, burning, and subsequent emissions. It is because of the consequences of the old, and the fact that it continues to dominate our global energy system, contributing disproportionately to current global CO_2 emissions, that carbon has rapidly become 'a common denominator for thinking about the organization of social life in relation to the environment'.[18] We see this in the emergence of vast political networks, institutions, and frameworks, from the local through to the global, dedicated to carbon's management and reduction, and in the emergence of a 'new carbon economy' through which carbon's commodification becomes a dominant strategy for organizing, amongst other things, climate change mitigation and adaptation. The remainder of this chapter will consider the nature of the international climate change regime that has developed since the early 1990s, including the role played by state and non-state actors,[19] paving the way for chapter 3's focus on the new carbon economy as it relates to mitigation, evident in the rise of carbon trading across the globe.

A history of global climate change politics
Beginning with the formation of the UNFCCC in 1992, through to the Kyoto Protocol of 1997, and on to the Paris Agreement of 2015, with the UNFCCC hosting a global Conference of the Parties (COP) to the Convention each year since 1995 to try to hammer out roadmaps for global and national emissions reductions, we see that 'historic responsibility' has remained immensely consequential to the evolution of governance frameworks. Its role has been embodied in the concept of 'common but differentiated responsibilities and respective capabilities' (CBDR-RC), which served broadly to distinguish between so-called developed and developing countries – or Annex I and non-Annex I countries as they came to be known respectively in UN negotiations. The principle of CBDR-RC, while recognizing that we all have a common responsibility to protect the planet's critical life-support systems that are currently threatened by anthropogenic climate change, acknowledges the much greater role played by the world's advanced industrial nations in contributing to the problem in the first place. Moreover, it accounts for vast differences in the capability to act on the problem, whether financial, technological, or otherwise, highlighting the capability of wealthier nations to do more. Given that superior capability has for many nations been the end result of a long his-

tory of uneven fossil-fuelled development, the argument can be made that wealthier nations have a responsibility to transfer some of that wealth to poorer nations to support low-carbon development, particularly when poorer nations are being told that the high-carbon path to development charted by the North has now been washed out, and is no longer an option for nations whose populations remain unacceptably poor.

While the battles over what place the principle of CBDR-RC should have in global governance frameworks have been particularly intense, developing nations were successful in having the text of the UNFCCC include language stating that Annex I countries 'should take the lead' in the fight against climate change and its impacts, with the Berlin Mandate from COP 1, held in Berlin, in 1995, cementing the principle further by setting out a process through which Annex I countries would take on emissions reduction targets globally, while exempting non-Annex I countries from any such commitments. The stage was thus set for negotiating the Kyoto Protocol of 1997 – the first agreement of its kind to set out legally binding emissions reduction targets for developed nations. These targets amounted to a collective commitment to reduce emissions to 5.2 per cent below 1990 levels between 2008 and 2012, the Protocol's first commitment period. The North–South divide was so contentious in the United States, then the world's highest-emitting nation, that national legislation was passed in the US Senate to bar the government from signing any global agreement, including the Kyoto Protocol, that would require the country to lower emissions without requiring the same of developing nations. Known as the Byrd–Hagel Resolution of 1997, this piece of legislation ensured that the United States, while highly vocal in global climate governance forums, would exercise its voice as an outsider that agitated for a common and undifferentiated approach to global emissions reductions. With the possibility of Kyoto's ratification by the USA lying in ruins – while it was signed on to under President Bill Clinton, President George W. Bush officially pulled out in 2002 – the treaty was left with 193 Parties made up of rich nations, those with economies in transition, and those in the global South, covering a fraction of global emissions. In 2011, the Conservative government of Stephen Harper in Canada withdrew from the Protocol, citing the absence of developing nation commitments in the global agreement, including from heavy emitters like China and India.

Such rhetoric intentionally underplayed, if not altogether ignored, the fact that the Kyoto Protocol represented a limited first step on global climate action intended to allow those predominantly responsible for global warming to show the rest of the world that they were serious about changing course. With only four years from start to finish, the Protocol was never designed for permanence, even though this is how its critics often framed it. Of course, it would be a mistake to assume that, had the principle of CBDR-RC not been an issue in global negotiations, meaningful and effective targets would have been set, since, at its core, real action on climate change requires that we challenge an economic system whose primary imperative is that of growth, thus sustaining our continued dependence on fossil fuels. Indeed, CBDR-RC, while highlighting the very real history of responsibility, has too often provided a convenient distraction, allowing powerful leaders to blame inaction on an unfair global climate regime rather than their desire to maintain their current status within the global international political economic system.

That the issue of historic responsibility would prove a major stumbling block in global negotiations was hardly surprising. As nations began to negotiate a successor to the Kyoto Protocol that could come into effect in 2013, the North–South divide remained one of the most contentious issues moving forward, alongside distributional conflicts over emissions reduction commitments. In terms of the former, China's ascendancy to become the highest-emitting nation globally underscored the shifting geopolitical sands upon which the Kyoto Protocol had been built. It also signalled a profound transformation in traditional alliances amongst Southern nations. For countries grouped within the Alliance of Small Island States (AOSIS), which negotiate as a block at UN climate meetings, anthropogenic climate change represents an existential threat. For especially low-lying small island developing states facing near-certain disappearance with rising sea levels, and for many of the world's forty-eight least developed countries (LDCs), all of whose past and present contributions to global GHG emissions are negligible, the idea that the world's emerging economies should be freed from stringent and legally binding targets, given their late arrival in the high-emitters' club, was unacceptable.

In terms of distributional conflicts, negotiating a successor to the Kyoto Protocol that would include legally binding targets was proving difficult, especially given the intense disagreements over burden-sharing that existed within and between the world's heavy emitters. It was at COP 15 in Copenhagen in 2009 that these major issues were revealed for the roadblocks that they were. Instead of delivering a global agreement to address the many challenges of global climate change and that could pick up where Kyoto was leaving off, Copenhagen ended in failure, dealing a massive blow to the global climate regime. This failure hinted at the possibility that securing legally binding emissions reduction targets and a clear distinction between North and

South in a future agreement might not happen. The Kyoto Protocol itself was extended for a second commitment period to fill in until a new agreement could come into effect, but its membership and effectiveness, for all their flaws, were diminished further when Russia and Japan announced that they would no longer remain parties to the Agreement.

Between Copenhagen's failure and late 2015, the world's nations negotiated feverishly to draft a new global climate agreement. The Copenhagen Accord of 2009 was a non-legally binding political agreement negotiated between a small group of national leaders, including the United States, and Brazil, South Africa, India, and China – or the BASIC countries – amongst others. Because of this, the Accord was criticized for side-stepping formal negotiations and reflecting the interests of the world's most powerful states. As such, delegates at COP 15 'took note' of the Accord, since consensus, required to adopt decisions, was not possible. The Accord set a 2015 target for review, while also endorsing a system of voluntary pledges from all nations, North and South. The groundwork was thus laid for the Paris Agreement of 2015.

The Paris Climate Agreement
The build-up to COP 21 in Paris, running from 30 November to 12 December 2015, was immense. While the Copenhagen COP failed to produce a successor to the Kyoto Protocol, the scientific data on catastrophic climate change continued to build, with many framing Paris as a last-ditch, critical moment in history to try to get it right. As the Paris COP drew to a close it was clear that a global Agreement would be forthcoming, although assessments of the final text range from scathing critique and despair, to cautious optimism, to celebratory embrace. There are a number of

key features that distinguish the Paris Agreement from its predecessor, and from which the wide gulf in assessments have emerged. Perhaps most importantly, the Agreement has been described as a bottom-up approach to global mitigation efforts since it is built on voluntary emissions reduction pledges. In the run-up to Paris, these came in the form of Intended Nationally Determined Contributions (INDCs) outlining a country's intended mitigation actions under the Paris Agreement, later to become Nationally Determined Contributions (NDCs) following formal ratification by member states. Equally important, Kyoto's hard distinction between North and South, embodied in the concept of CBDR-RC, was softened considerably. Now, mitigation efforts would be assessed on a spectrum, with all countries participating. At one end, advanced industrial nations are to continue pursuing 'economy-wide absolute emission reductions targets'; from there, developing nations are able to determine where they position themselves, while committing to the continued enhancement of their 'mitigation efforts'.

While the text of the Agreement states that global temperature increase should be kept 'well below' 2 °C of warming, small island states grouped under AOSIS, along with their allies, were successful in having the aspirational limit of 1.5 °C included in the text, a limit much more consistent with acknowledging the threats they face. Developing countries also succeeded in having loss and damage included in the Agreement under Article 8. While Article 8 elevates the issue of loss and damage and the recognition of the adverse and devastating impacts climate change is having on the world's poorer nations – this building from the creation of the Warsaw International Mechanism for Loss and Damage that came out of COP 19 in Warsaw, Poland, in 2013 – it remains severely limited. In particular, at the insistence of richer-country governments in their desire to ensure that poorer nations would have no basis for claiming legal and financial liability for the adverse effects of climate change, Article 8 notes that 'it does not involve or provide a basis for any liability or compensation'.

Such wording foreshadows the content of Article 9 of the Agreement which deals with climate finance to the global South. Again, the Article contains no precise financial commitments, instead calling on developed country Parties, who 'should continue to take the lead in mobilizing climate finance from a wide variety of sources, instruments and channels, noting the significant role of public funds'. It continues that 'such mobilization of climate finance should represent a progression beyond previous efforts', while encouraging other Parties 'to provide or continue to provide such support voluntarily'. By allowing developed countries to *mobilize* finance from 'a wide variety of sources', the Paris Agreement blunted Southern calls for firm commitments to new and additional financial resources from the North. Reference to the maintenance of the goal of $100 billion per year in climate adaptation and mitigation finance through to 2025, a figure negotiated in 2009 in Copenhagen, was ushered out of the main legal text of the Agreement and into the 'Decision document'. The Decision document is not legally binding, with observers noting that, had $100 billion been included in the legal text of the Paris Agreement, Barack Obama would have been required to get approval from Congress in the United States. Similarly, Congressional approval would have been required had binding targets been included in the main text, with many highlighting the specific intentionality of the Agreement's legal architecture: it ensured that it would never reach a hostile US Congress, allowing instead for ratification by executive decree.

In all, there are 197 Parties to the Paris Agreement, with 170 having ratified the agreement at the time of writing. In order to enter into force, 55 Parties to the Agreement, accounting for at least 55 per cent of global GHG emissions, were required for ratification. As the Paris COP concluded, negotiators were hoping that these thresholds would be reached so that the Agreement would be operational by 2020, the year it is scheduled to take effect, yet, less than one year after the Paris COP, on 4 November 2016, the Agreement entered into force. The speed with which countries moved to ratify the Agreement suggests that climate change has finally 'arrived'; it is a global threat that can no longer be ignored. At the same time, however, it is a four-year process to withdraw from the Agreement, meaning that the Obama Administration and ally nations were anxious to bring it into force as quickly as possible, given the possibility of a Trump victory in the US election of 2016. As the world now knows, Trump did win the Presidency in the USA, getting to work immediately to undo Obama's climate change policies domestically – indeed, decades of environmental policy in general. Now that the Trump Administration has confirmed that it will withdraw from the Agreement, we can be certain that efforts to meet Obama's climate targets delivered in Paris are now dead.

Moreover, critics of the Paris Agreement highlight the fact that it is made up of voluntary national commitments and no overarching and legally binding target for emissions reductions, meaning it lacks teeth and thus incentives for states to follow through on their NDCs. Many point to the studies that analyse the effectiveness of current NDCs – when taken together, they suggest we are on a path to see 2.7–3.5 °C of warming, or more – an absolutely catastrophic outcome that will hit the poorest and most vulnerable hardest and fastest. Given the magnitude of the transition required to stay below 2 °C, let alone 1.5 °C, alongside the powerful and deeply entrenched economic and political interests that continue to resist this transition, many believe that the voluntary and non-punitive nature of the Agreement that allows states to determine their targets independently, and whether or not they will follow through, is nothing short of disastrous. Others, equally critical of the Agreement's lack of ambition, recognize that, had the Paris COP demanded absolute and legally binding emissions reduction targets, another failure was inevitable, suggesting that, sadly, this was the very best they could get. In order to address this lack of ambition, the Agreement itself lays out a legally binding framework to allow for dynamic improvements. Article 4 of the Agreement requires states to submit new NDCs every five years, stipulating that each new NDC must 'represent a progression beyond the Party's then current nationally determined contribution and reflect its highest possible ambition'. Moreover, Article 14 requires a process of regular 'global stocktakes' to review 'collective progress' in achieving the goals of the Agreement, with the first to take place in 2023 and then every five years. Referred to as the Agreement's 'ratchet mechanism' in that it provides Parties with the opportunity to ratchet up their ambition at regular intervals, the requirement for global stocktaking and the submission of new and increasingly ambitious NDCs every five years gives hope to some that, while imperfect, the Agreement can be strengthened to achieve meaningful climate ends.

Time will tell what the Paris Agreement will deliver. Although the Agreement is critically important, as a creature of the UNFCCC it represents but one element in the broader global climate governance framework. Observers have for many years been tracing the increasing importance of subnational state and non-state actors in the global climate regime, noting that the Paris Agreement for the first time recognizes stakeholders not Party to the Agreement as key players in achieving its ends. These actors comprise a wide array of interests and strategies, demonstrating the substantial diffusion of carbon governance and management processes beyond the UNFCCC. Some remain state-based, evident in a long list of climate actions and initiatives hosted by subnational state actors.[20] We see this in transnational city networks involved in climate action – including the C40 Cities Climate Leadership Group that brings together over 90 of the world's largest cities to cooperate on climate action – and in the Under2 Coalition that brings together 205 subnational governments, including states, provinces, and cities from around the world, committed to lowering their emissions 80–95 per cent below 1990 levels by 2050, while cooperating on climate action. Other initiatives combine state and non-state actors in hybrid forms of governance. The recently announced Global Alliance to Power Past Coal – a coalition of 26 countries, 8 subnational governments, and 24 businesses and organizations that have committed to work collectively to phase out coal – and the Carbon Pricing Leadership Coalition (CPLC) – which includes 32 national and subnational governments, roughly 150 private businesses, and many more industry associations, NGOs, global institutions, and civil society actors that have joined together to support the development of carbon pricing around the world – represent this trend. We Mean Business, a non-profit coalition that works on carbon management with close to 650 businesses around the world, and the International Emissions Trading Association (IETA), a global business organization representing industry and finance that lobbies for the adoption of carbon markets around the world, are examples of private-sector actors mobilizing to participate in and shape climate governance and action. Formerly known as the Carbon Disclosure Project, the CDP gathers data voluntarily disclosed by companies and various levels of government around the world, tracking their environmental performance and carbon impact. Their reports are now widely used and cited, advancing the transparency agenda in global carbon governance.

Through these examples, representing but a fraction of governance strategies beyond the formal proceedings of the UNFCCC, we see cooperation on information and knowledge sharing and diffusion, data collection, technology and innovation, carbon finance, and much more. The extent to which the private governance of climate change – through strategies such as carbon audits and accounting, certification schemes, shadow carbon pricing at the company level, labelling initiatives, and so forth – intermingles with the public is now significant. In some cases, we are seeing important advances in pushing forward a low-carbon agenda. Some state and institutional investors are noteworthy here as one set of actors pushing for fossil fuel divestment, given the threat of stranded assets in the fossil fuel sector as the world moves to a lower-carbon future. For pension funds that have a fiduciary duty to invest responsibly, it is becoming increasingly difficult to justify investments in the global fossil fuel sector.

Referred to by some as 'hybrid multilateralism', we see today a climate governance landscape influenced deeply by a wide range of actors, both state and non-state. It remains vitally important, however, to analyse the power relations that influence these processes, whereby some actors are able to leverage their financial, economic, and thus political power to ensure a strong presence and voice in governance efforts globally, thereby helping to shape how they take form.

Conclusion

The lessons of Paris are instructive for thinking through the global political economy of carbon. In global bodies like the World Trade Organization, and bilateral and multilateral trade agreements, we see how our leaders and the world's business community can come together and hammer out agreements with teeth, to be used against those who choose to break the rules. Legally binding, dispute resolution and arbitration mechanisms, protecting investor rights – these agreements reveal the extent to which the world's elite are willing to go to protect the global economic system in its current form, and the powerful economic actors within it, often in ways that run counter to addressing global climate change. That a similar level of commitment and protection for the planet and the world's people – for the most vulnerable amongst us, for whom climate change represents another layer of oppression that amplifies already-existing class, racial, and gendered forms of oppression – was not forthcoming in Paris, should hardly surprise us. Unlike trade agreements that seek to extend the current global political economic system, addressing climate change requires that we confront that system and the fossil capital so deeply embedded within it. At its core, the issue of carbon is an issue of power all the way down. In acknowledging this, along with the insights provided in this chapter on carbon's global political economy, we are well situated to refine our gaze for the next chapter. Taking us deeper into the world of global and national climate policy, chapter 3 explores the rise and increasing dominance of carbon trading globally, an approach to mitigation through which carbon's commodification is supposed to unleash deep decarbonization. Given that it is the global political economy of carbon and fossil capitalism that have given birth to carbon trading, since it poses little real disruption to business as usual, there is significant cause for concern. It is to this topic that we now turn.

CHAPTER THREE

Trading Carbon to Cool the World?

This is a chapter about carbon trading, by far one of the most popular and expanding policy frameworks being used to fight climate change globally. Carbon trading, which includes carbon offsetting, is nothing if not highly complex. Within the world of the carbon trade and the market that supports it, myriad issues, challenges, and uncertainties collide, which are being amplified as negotiators and policy makers now seek to build on the market mechanisms of the Kyoto Protocol to construct a global carbon market within the framework of Article 6 of the Paris Climate Agreement. All of this matters in profound ways, given the weight of the climate challenge that confronts us. As the writing for this chapter began, a news article began circulating on social media titled 'No Country on Earth Taking the 2 °C Target Seriously', followed days later by another: 'Earth Could Break Through a Major Climate Threshold in the Next 15 Years, Scientists Warn'. Both summarize peer-reviewed scientific data suggesting that we are likely to blow through the climate targets set by the international community, with the latter reporting on recent research that predicts we could pass the 1.5 °C threshold of warming above pre-industrial levels in as little as fifteen years. And yet we continue to build and expand fossil fuel infrastructure, subsidizing an industry whose known reserves put us well past 2° C of warming. So the question emerges: can carbon trading pull us back from the brink? Does it have what it takes to respond effectively to the urgency of the crisis upon us, and is this what drives its growing popularity amongst policy makers, government leaders, much of the global business community, and certain environmental non-governmental organizations (ENGOs)? Perhaps a few examples can help to set the stage for the discussion that will follow, underscoring those issues that are ever-present in the debate on carbon trading, though little acknowledged by many of its proponents. Let's start with satire.

It was in 2007 when a video started circulating on the internet, sent and re-sent via email, posted on listservs, and shown in classrooms, to colleagues, and to friends. The promotional video for a new company, CheatNeutral.com, promised that for a small fee – a mere £2.50 – clients could offset their infidelity. Specifically, individuals who had cheated on their significant other would now have an easy way to 'neutralize the damage done' to their relationships – they could pay £2.50 to CheatNeutral.com who would then pay individuals or couples who remained faithful in their relationships, thereby ensuring that 'total levels of heartbreak in the world' did not go up. A first in its class, it would seem that CheatNeutral.com had come up with an innovative and cost-effective strategy for dealing with one of life's most difficult situations. Absurd? In a nutshell, yes, and this was the point. Indeed, the creative team behind Cheatneutral.com harnessed the seemingly outrageous to draw attention to the little-known but rapidly growing market in carbon offsetting globally – a market that allows emitters to continue emitting, while paying someone else, somewhere else, to lower their own carbon emissions in theory, or to avoid future carbon emissions. Especially clever, Cheatneutral.com used parody to prompt its viewers to take a close look at this emerging approach to climate policy, within which mitigation strategies too often distract from the task at hand: the rapid reduction of carbon emissions globally in order to avert sweepingly catastrophic climate change.

Entirely without humor, on the other hand, we have the real-life example of PopOffsets, a project run by the UK charity Population Matters. According to the PopOffsets website, for a modest payment – roughly €64 for a resident of the European Union and $140 for a resident of the United States, based on average annual per capita carbon emissions – an individual can offset a year's worth of emissions, with PopOffsets offering grants to projects that work to provide family planning services globally. The site links to an online carbon calculator that allows purchasers to calculate their annual emissions in order to determine the appropriate payment level, opening the door for organizations and companies to participate as well. In short, PopOffsets allows you, me, or anyone else, to pay so that women might not have babies. A good news story for those concerned that action on climate change might require changes in lifestyle, or productive and consumptive activities. To be sure, the team at PopOffsets think they've identified the ultimate 'win-win' strategy in population offsetting, stating on their website: 'after all, an absent human cannot emit carbon dioxide', and nor can their 'nonexistent descendants'.[1]

PopOffsets provides an ideal starting point for this chapter since it gets to the heart of a number of issues that remain troublingly absent from discussions on carbon trading and offsetting, particularly amongst their proponents, who tend to frame both as relatively easy technical fixes for one of the greatest challenges facing contemporary society. While we can laugh at the premise underpinning CheatNeutral.com, certain that a simple monetary transaction provides no real absolution for the betrayal that comes with infidelity, in today's 'green economy' the opposite becomes true. Through population offsetting, Jack pays a fee so that Jane doesn't have a baby, allowing Jack to continue driving his SUV. Rather than exploring low-carbon alternatives, Jack can harness his purchasing power to lower his emissions, *money* and *action* now functionally equivalent. Except they aren't, since Jack didn't actually lower his carbon emissions, not to mention the fact that there is no credible way of measuring the equivalency between Jack's emissions and those of the non-existent human being. This is an unsettling realization for many concerned about the urgency of the climate crisis and the magnitude of change that is required. Equally troubling, though perhaps less visible to some, are the structures of oppression upon which something like population offsetting, and carbon trading and offsetting in general, rest, for they assume what we might call *unproblematic commensurability*.

Things are commensurable when they can be measured by a common standard. Unproblematic commensurability occurs when it is assumed that, in the case of carbon, it's all the same: no matter who is emitting, or where, how, and why, or who is reducing, or where, how, and why, there is no difference. This assumption reduces carbon dioxide – CO_2 – to its molecular structure and its heat-trapping properties, devoid of the specific historical, political, and social context within which it is embedded, or the power relations through which it is shaped. The simple goal: emit less CO_2. But is it really all the same? By examining how the problem that PopOffsets seeks to respond to is framed, we see how unproblematic commensurability operates and why this question matters.

The PopOffsets website states that 'one of the most cost-effective ways of minimizing climate change' is to help people to avoid 'unplanned pregnancies'. Suggesting an 'intrinsic link' between global population growth and climate change (more people = more carbon), the message prioritizes human reproduction and unchecked birthrates as perhaps the greatest threat to planetary survival. The language used to convey their message frames overpopulation as a global problem, with the site highlighting their work in both developed and developing countries to reduce birthrates. Nevertheless, it is through the photos used on the website to construct the story of PopOffsets visually that a particular geography and politics of space is revealed – classed, racialized, and gendered – which is premised on a firm divide between developed and developing, rich and poor. It is thus that anyone browsing the site will encounter multiple photos suggesting that the target of family planning services is largely the stereotypical 'third-world woman' – dark-skinned, poor, and fertile. She cooks in a poor dwelling with her baby on her back somewhere in Latin America, we think; she seeks shelter from the sun under a tree in a parched landscape somewhere in Africa; she squats barefoot on the ground descaling fish in India, or maybe Bangladesh, the file name for the photo identifying her only as the 'impoverished woman'. Images of an overcrowded city, another heavily polluted and thick with deadly smog, and a slum on a hillside, suggest that these are the problems of developing countries. Through visual representation, an overpopulated third world, along with the bodies of its most marginalized women, become the implicit targets of climate action, with the site constructing a corresponding narrative around who should do the intervening. Click the 'Offset Carbon' tab – it is here where one can purchase offsets – and one finds pictures of a passenger jet and a luxury car, polished and shining gold in colour. Who flies on these planes and drives such cars, for whom PopOffsetting as mitigation strategy is supposed to respond? Click the 'Contact' tab to see who the typical offsetter might be – a white woman, blond-haired and blue-eyed, using her mobile phone to get in touch.

They say a picture is worth 1,000 words. The tale told by these pictures spatializes responsibility in highly troubling ways. The narrative casts poor brown women as a major threat to planetary survival. That these women and their children bear no real present-day, or historical, responsibility for climate change matters not. That the average American, Canadian, or Australian has a per capita carbon footprint over twelve times that of the average Bangladeshi citizen, and even greater compared to that of poor Bangladeshi women, is unimportant. While acknowledging that poor brown women are not super-consumers, it is implied that they, and their offspring, want to be, which is quite possibly the greatest threat of all since clearly there are few to no seats left at the table. And let's be clear, those already seated will not be asked to give anything up. The problem with offsetting considered this way is that it insulates the privilege of the super-producer, the super-consumer, and the super-emitter, whether residing in the North or South, in deeply classed, racialized, and gendered ways, displacing the burden of responsibility onto distant *others* now signified as threats.

To be fair, PopOffsets is probably the only organization of its kind – its offsets are not certified in any compliance or voluntary carbon markets globally, providing instead a niche opportunity for those especially concerned with issues of population and climate change. Nevertheless, the logic of population offsetting emerges directly out of what is taking place in actual carbon markets around the world. Carbon trading and offsetting are justified with reference to the *flexibility* they provide to emitters in achieving their emissions reduction targets. So the factory, or firm, or individual, finding it difficult to lower their carbon emissions at the source, now has the flexibility to trade and offset their carbon, allowing them to claim actual emissions reductions as their own. In this it seems that 'flexibility' functions as a form of Newspeak, the fictional language of Oceania in George Orwell's *Nineteen Eighty-Four*, which was 'devised to meet the ideological needs' of those in power, while 'diminish[ing] the range of thought'.[2]

Within an economic system that requires continual growth and that remains predominantly driven by carbon-intensive fossil fuels, the economic cost of rapid emissions reductions at the source poses an existential threat to that system, the power relations underpinning it, and its many dependencies. Flexibility via financial transaction thus becomes synonymous with one having actually reduced one's own carbon footprint. Comfortable now that something is being done, we need not think beyond these boundaries.

By 'diminishing our range of thought', carbon trading confines the questions we should ask to those of calculation and accounting – how can we measure the precision of the trade; do we have a tonne of CO_2 here for a tonne of CO_2 there? The questions masked by assumptions of unproblematic commensurability – those of environmental effectiveness, justice, and power – move beyond the range of the frame. Questions concerning how carbon trading extends the reach of the capitalist market and its growth imperative into ever more areas of planetary life – whereby CO_2 and associated biophysical processes become commodities reduced to their market value, assessed on their potential to spur economic growth while earning their keep – fall similarly outside the range of the frame.

Building on chapter 2's exploration of the global political economy of carbon, this chapter takes the reader on a tour of carbon trading and carbon markets globally, charting their ascendance, even in the face of crisis and collapse, to become one of the most popular climate change mitigation strategies in use today. Borrowing in part on the important distinction between 'problem-solving theory' and 'critical theory' made by Robert Cox in his seminal 1981 contribution to International Relations theorizing,[3] it is argued that carbon trading must be understood as a 'problem-solving' approach to dealing with climate change, as its proponents both normalize and naturalize the current global political economy of fossil capitalism, with consequent efforts to deal with climate change intended to safeguard the smooth functioning of that system. In this vein, climate change is understood as a 'particular source of trouble', with carbon trading an attempt to minimize the impact of this trouble, namely the threat it poses to profitability. It is thus that 'flexibility' promises minimal disruption to business as usual, allowing us to appreciate in part the current appeal of carbon trading, in addition to the fact that it offers new opportunities for growth and accumulation with the expansion of markets in carbon. Stating these facts should not be controversial. As the previous chapters have shown, the interconnected interests and dependencies that are the hallmark of our current global economic system mean that it is incredibly difficult to take seriously, and respond adequately to, a problem that is so all-encompassing and that strikes at the very heart of that system. While we do have choices, and alternatives do exist, many of which will be discussed in much more detail in the next chapters, the political economy of climate change mitigation is one marked by struggles over power in order to control the terrain upon which climate action takes place. In this sense, by fortifying and safeguarding the current world order, carbon trading represents a victory for those who benefit most from that order. It is within the tradition of Cox's critical theory and its call for emancipatory practice and an articulation of alternative orders that this chapter frames carbon trading. With that as our reference point, we can move now to consider the content, history, complexities, and contradictions that comprise the world of global carbon trading.

What is carbon trading?

Before considering the history of carbon trading, a basic primer is in order. Whether it's called carbon trading, emissions trading, or cap-and-trade, carbon trading is part of a movement to put a price on carbon. This movement has grown over the last few years, with proponents arguing that it is necessary since carbon dioxide represents what economists call a 'negative externality' – a by-product of the day-to-day operation of the market economy that isn't costed into economic transactions and thus isn't accounted for. In this way, economists note that the externality is free – if a factory spews CO_2, warming the planet and leading to drought that devastates crops in Uganda, it is the farmers and consumers who pay, but not the factory owner. By pricing carbon, it is argued, the negative externality is brought into the economic equation, making it more expensive to emit CO_2. The idea is that, as CO_2 gets more expensive, emitters will have an incentive to look for low-carbon alternatives, spurring the low-carbon transition that the economy needs.

There are two main policy instruments dominating discussions on carbon pricing around the world.[4] The first is a carbon tax, under which a government sets the tax level, or price, per tonne of CO_2 equivalent (tCO_2e), requiring covered emitters to pay the tax based on emissions levels.

Governments must choose which sectors to include under the tax, but once those details are worked out it operates in a fairly simple way. Worldwide, there are at least twenty-one national and two subnational jurisdictions that are taxing carbon, with an estimated combined value of US$19.39 billion in 2017. Some of the highest carbon taxes per tCO_2e globally include Sweden's at $126, Switzerland's at $84, and Norway's at $52. At the low end, Mexico, Ukraine, and Poland all have taxes of less than $1 per tCO_2e.

The second policy instrument, carbon trading, entails a government setting a cap on allowable emissions within a jurisdiction over a specified period of time. Covered emitters, or compliance entities, then have a quantitative limit on their allowable emissions under the cap, something that should progressively decline over time, with the government creating carbon allowances equivalent to the cap, each worth 1 tCO_2e. In simplified terms, if a government caps emissions at 100,000 tCO_2e for a given year, an equivalent 100,000 carbon allowances will be created, then to be auctioned or, as is often the case, distributed for free to compliance entities. Cap-and-trade is thus a quantity-based mechanism for dealing with negative externalities. While not direct carbon pricing as in taxation, carbon trading tends to be treated as a pricing mechanism since one of its core functions is to create a market within which allowances can be bought and sold, with price determined through the interplay of supply and demand relative to a progressive lowering of the cap. According to the theory of cap-and-trade, if an emitter finds it relatively easy to lower its emissions, it will have excess carbon allowances to sell, or trade, to firms that experience greater difficulty. Fossil fuel companies and carbon-intensive firms producing cement and steel are examples of those that face greater costs and difficulties in lowering their emissions. Within

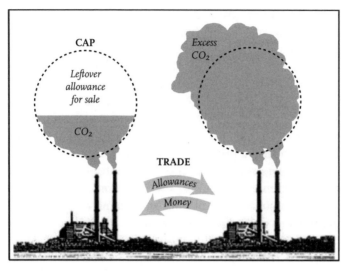

Figure 3.1 Cap-and-trade in theory

Source: McCormick, www.mccormick.northwestern.edu /magazine/spring-2013/carbon-calculator.html.

a carbon market, these firms can meet their compliance obligations by purchasing carbon allowances, which proponents cite as a double economic advantage. First, they note that trading provides heavy emitters with flexibility in meeting their targets since trading is cheaper than the investments that would be needed to achieve comparable emissions reductions at the source. The economic health of these sectors is thus maintained. Second, they argue cap-and-trade provides an economic win-win for firms with lower abatement costs since emissions cuts save them money, with excess carbon allowances then providing a new revenue stream once sold to heavy emitters. Figure 3.1 illustrates the theory of how cap-and-trade is supposed to work.

In the mainstream policy debate over carbon pricing, proponents of trading argue that it is the most cost-effective strategy for lowering emissions since it is the 'invisible hand' of the market that determines where emissions reductions can be achieved most efficiently and thus at lowest cost. Taxation is criticized for what market advocates, particularly those inspired by a neoliberal logic, perceive as arbitrary price-setting by inefficient government bureaucracies – a process sure to penalize economic activity and stifle growth. It is at this juncture – that between effective climate action and the need for continued economic growth, with cap-and-trade intended to bring the two together – that the nature of emissions trading, and the political economy that shapes it, is best understood. In its 2017 report, the High-Level Commission on Carbon Prices, chaired by economists Joseph E. Stiglitz and Nicholas Stern, suggests that 'climate policies, if well designed and implemented, are consistent with growth', and that 'the transition to a low-carbon economy is potentially a powerful, attractive, and sustainable growth story'. Fitting with the growth imperative of the current economic system, this narrative acts to erase the otherwise obvious contradictions between the needs of the economy and those of the planet – with the right climate policies, including a price on carbon, we can have our cake and eat it too. Enframed thus, the issue of limits, or the idea that what we actually need is less production, less consumption, and less growth, is effectively externalized beyond the bounds of the problem. So how did we get here?

Carbon trading: a history

With the adoption of the Kyoto Protocol in 1997, carbon trading, including cap-and-trade and carbon offsetting, was formalized in international policy and law, with the first regional cap-and-trade scheme starting in 2005 with the European Union Emissions Trading System (EU ETS).[5] Since then cap-and-trade schemes have spread regionally, nationally, and subnationally across the globe. At the time of writing, there were 19 cap-and-trade schemes operating worldwide in around 40 countries, with the number of countries participating in carbon trading ballooning to over 100 when carbon offsetting, primarily under the Kyoto Protocol's Clean Development Mechanism (CDM), is included.

The intellectual case for emissions trading extends back to the 1960s and 1970s. During that time, the field of economics was increasingly valorized as a neutral science that could inform all facets of public policy, environmental economics as a sub-field was increasingly institutionalized, and a debate on the use of taxes versus market mechanisms to deal with environmental harm became particularly active. Free market economists from the neoclassical tradition argued, drawing especially on the work of Ronald Coase and his 1960 article 'The Problem of Social Cost', that the market could best achieve environmental ends at the lowest cost and greatest efficiency, with this argument then used to push for emissions trading to deal with a host of problems. Support for emissions trading was also situated within a broader shift to neoliberal forms of governance in the 1970s and 1980s both in the USA and internationally, which sought to supplant more directly regulatory and Keynesian-inspired approaches to governance with a free market ethos that persists to this day.

The first experiments in emissions trading took place in the United States and were limited to offsetting, or baseline-and-credit, in order to provide emitters with flexibility in meeting regulatory requirements coming out of the Environmental Protection Agency (EPA) and the Clean Air Act (CAA) of 1970. With the CAA Amendment of 1990 under President George Bush Sr., a fully operational emissions trading scheme was planned, coming into effect in 1995 with the new market in sulphur dioxide (SO_2). Addressing in part the problem of acid rain resulting from SO_2 emissions largely from coal-fired electricity generation, this experiment has been touted as wildly successful, though evidence now shows that much of its cost-effectiveness came not from the existence of the market itself, but from the deregulation of rail rates during the 1970s and 1980s, and to lower abatement-technology costs than initially anticipated. Moreover, data published in 2009 showed that cap-and-trade had failed to 'sufficiently stabilize' acid rain, with 'forest decline' continuing (p. 104).[6] Other experiments in emissions trading would spread throughout the USA in the 1990s, with Simons and Voß noting that an important 'emissions trading constituency' had formed that carried with it significant 'epistemic influence on national and transnational policy making processes' (p. 56). Paterson highlights the growing influence of this community during the 1990s, with the formation of various emissions trading networks that brought together well-known economists writing on the subject and business and state actors in the UK, USA, and EU.[7]

Indeed, as a nascent global climate regime was taking shape, starting in part with the creation of the IPCC in 1988, followed by the UNFCCC in 1992 with its Conferences of the Parties (COPs) that would convene annually starting with the Berlin COP in 1995, the influence of this growing epistemic community would prove instrumental in shaping the Kyoto Protocol of 1997. A central player in this community was the American government under the Administration of Democratic President Bill Clinton. Both

Clinton and his Vice President, Al Gore, were staunch advocates of free market approaches to environmental governance, with Clinton famously arguing that, through a market-based approach to climate change mitigation, there would be 'not costs, but profits, not burdens but benefits, not sacrifice but a higher standard of living' (quoted in Meckling, *Carbon Coalitions*, p. 87). The USA thus made its participation in a global climate treaty contingent on the inclusion of emissions trading as an option for states to comply flexibly with GHG reduction targets.

Other key participants in the growing epistemic community included members of an emerging business coalition favouring emissions trading, including oil majors like British Petroleum. In the early days of global climate change negotiations, many carbon-intensive businesses opposed the development of an international climate agreement, citing the impact it would have on their profitability. The Washington DC-based Global Climate Coalition (GCC) epitomized this position, with its members from the fossil fuel, chemicals, auto, and steel sectors, amongst others, aggressively resisting global efforts to regulate carbon. Corporate unity didn't last, however. Various business interests threw their support behind emissions trading for a number of reasons. First, many began to appreciate what their absence at the negotiating table would mean: the absence of their voice in shaping an agreement that would best suit their economic interests. In this case, carbon trading promised minimal disruption to dominant relations of production and consumption, with a coalition of businesses in favour of trading mobilizing to advocate for it. Paterson notes that insurers and institutional investors also played an instrumental role in allowing carbon trading to gain political legitimacy.[8] Both are aware of the risks posed by climate change to their business models; equally important, various financial actors saw the business opportunity in the creation of new markets. While organizations were inspired by different concerns, the 1990s thus saw the emergence of strong support for carbon trading from the business and finance communities. Of course, some continued aggressively to oppose action on climate change, including companies like the US-based Exxon Mobil whose campaign to sow the seeds of climate change doubt and denial in the USA is well documented. The company's former CEO, Rex Tillerson, was appointed Secretary of State under Donald Trump.

ENGOs played a crucial role in this process as well, with the US-based Environmental Defense Fund (EDF) and the World Resources Institute (WRI) carrying out studies, writing reports, and lobbying aggressively in favour of emissions trading. Equally – if not more importantly – however, by working closely with those in the business community who favoured carbon trading, including fossil fuel companies that had been cast as the villains in the climate change drama, these ENGOs lent a degree of credibility to the villains, helping to recast them in shades of green. Given their cultural capital, supportive ENGOs helped to smooth the path for carbon trading as an *idea* in and of itself.

Opposition to carbon trading persisted, of course. In addition to certain sectors of the business community globally, many national governments in the EU and the global South, and many ENGOs and civil society groups, while supporting action on climate change, opposed carbon trading. Their opposition stemmed from an overall distrust of carbon trading as a tool that would simply allow rich countries – the super-emitters – to continue with their carbon-intensive growth. In response, the EU and many ENGOs advocated for a framework that would facilitate the meeting of emissions reduction targets via taxation and command-and-control regulations, supported by many Southern states that remained adamant in their calls for the North to assume its historic responsibility for the climate crisis by reducing its emissions at the source.

Nevertheless, the balance of power was with the USA, thus leading to acceptance of the terms it put on the table. The Kyoto Protocol was signed on 11 December 1997, in the Japanese city of Kyoto. It came into effect in February 2005 with 193 Parties – when Canada withdrew in 2011 under Conservative Prime Minister Stephen Harper, it was reduced to 192 – with its first compliance period running from 1 January 2008 to 31 December 2012. Parties to the Protocol were split between Annex B and non-Annex B countries, the former comprising developed countries with binding emissions reduction targets, and the latter including those in the global South without binding targets in line with the principle of CBDR-RC. Annex B countries were collectively required to reduce their emissions to 5.2 per cent below 1990 levels.

Carbon trading goes global

Known as Kyoto's flexible mechanisms, carbon trading within the Protocol included emissions trading, Joint Implementation (JI), and the CDM. The former, specified in Article 17 of the Protocol, allowed Annex B countries to trade Assigned Amount Units (AAUs) to meet their commitments. Specifically, their emissions reduction targets were expressed as allowable emissions during the 2008–12 commitment period, quantified then as assigned amounts. Countries surpassing their targets would be left with excess AAUs, each worth 1 tCO_2e, which could be sold, or traded, to countries unable to meet their targets. JI and the CDM are offsetting mechanisms. JI, detailed in Article 6 of the Protocol, allowed countries with emissions reduction targets, or their companies, to undertake offset projects jointly, producing Emissions Reduction Units (ERUs) worth 1 tCO_2e each that could be traded and counted towards the buyer's Kyoto target. The majority of JI took place in countries with economies in transition in the former Soviet Union. The CDM, the dominant source of carbon offsets within the Kyoto framework and outlined in Article 12, allowed countries without emissions reduction targets in the global South to undertake offset projects, earning carbon credits known as Certified Emission Reductions (CERs), worth 1 tCO_2e each, that could be sold to Annex B countries to count towards their targets.

In addition to Kyoto's flexible mechanisms at the international level, there are now nineteen emissions trading systems (ETSs) operating worldwide at the regional, national, and subnational levels. Table 3.1 provides details on each.

At the time of writing, existing ETSs covered 5 $GtCO_2e$, or 9.9 per cent of global GHG emissions, worth an estimated US$32.84 billion in 2017. Some were conceived within the context of the Kyoto Protocol to assist Parties in meeting their targets, including those in the EU and New Zealand. These markets, while regulating compliance entities within national and regional settings, were also fully integrated into the Kyoto framework, with JI and CDM carbon offsets playing a large role in each. Other ETSs were developed to support national or subnational emissions reduction targets outside of the Kyoto framework, including RGGI and California's Cap-and-Trade Program.

Today, there exists a patchwork of carbon trading schemes globally, with Article 6 of the Paris Agreement intended to stitch them together in a global carbon market. The inclusion of Article 6 in the Paris Agreement was a surprise for many, signalling to the world that carbon markets were being positioned to play an important role in

Table 3.1 Emissions trading schemes globally[9]		
Emissions trading schemes	Start year	Prices (US$) – 1 April 2017 (per tCO_2e)
Regional ETSs		
European Union Emissions Trading System (EU ETS)	2005	$4.99
National ETSs		
New Zealand ETS	2008	$12.02
Switzerland ETS	2008	$6.49
Kazakhstan ETS	2013	N/A
South Korea ETS	2015	$18.46
Subnational ETSs		
Regional Greenhouse Gas Initiative (RGGI)	2009	$3.45
Saitama ETS (Japan)	2011	$13.39
California Cap-and-Trade Program	2012	$13.81
Beijing Pilot ETS (China)	2013	$7.57
Guangdong Pilot ETS (China)	2013	$2.23
Quebec Cap-and-Trade Program	2013	$13.81
Shanghai Pilot ETS (China)	2013	$5.67
Shenzhen Pilot ETS (China)	2013	$5.08
Tianjin Pilot ETS (China)	2013	$1.97
Chongqing Pilot ETS (China)	2014	$0.87
Hubei Pilot ETS (China)	2014	$2.46
Fujian Pilot ETS (China)	2016	$5.02
Ontario	2017	$13.58
City ETSs		
Tokyo Cap-and-Trade Program	2010	$13.39

the post-Paris global climate architecture. On carbon markets, Article 6 lays out two important provisions: 6.2 and 6.4. Prefaced in Article 6.1 that acknowledges that some Parties may elect to 'pursue voluntary cooperation in the implementation of their nationally determined contribu- tions to allow for higher ambition', Article 6.2 refers to 'internationally transferred mitigation outcomes' (ITMOs) as the units, or carbon credits, that will be traded between Parties, while Article 6.4 establishes a 'mechanism to con- tribute to the mitigation of greenhouse gas emissions and support sustainable development'. Observers anticipate that this Sustainable Development Mechanism (SDM), as it is now widely referred to, will likely share some simi- larities with JI and the CDM, but will also reflect the fairly dramatic changes that have occurred in the international landscape of climate change policy, whereby all Parties to the Paris Agreement have committed to mitigation activi- ties in line with their NDCs. Specifically, while the CDM offered poorer nations the opportunity to sell carbon off- sets to those in the global North that were attempting to meet their emissions reduction commitments under Kyoto, these same nations will now be working to achieve their own emissions reduction commitments under the Paris Agreement. Also noteworthy is Article 5 of the Paris Agreement, which deals with carbon sinks and reservoirs – those areas that store and/or sequester carbon from the atmosphere, including oceans and forests. Article 5.1 notes that 'Parties should take action to conserve and enhance, as appropriate, sinks and reservoirs of greenhouse gases [. . .], including forests.' Article 5.2 goes on to note that this may be done through an array of approaches, including results-based payments, 'joint mitigation and adaptation approaches', and so forth. When read in conjunction with Article 6, we can see the framework that will underpin the potential inclusion of carbon forestry, discussed below, in a global carbon market.

Analysing existing and future carbon markets reveals how deeply complex and opaque they are, with a fundamental critique of these markets targeting this complexity on at least two fronts. The first centres on the sheer complexity that goes into making and running these markets, leaving few who actually understand emissions trading and how it works. Trading requires the commensuration of vastly different GHGs situated within vastly different socio-historical and ecological contexts into one tradeable commodity – the tCO_2e; the creation of corresponding systems of accounting and measuring, reporting, and verification (MRV) to track emissions, emissions reductions, and trades, and to ensure that reductions aren't double-counted by selling and buying nations; and the development of an elaborate regulatory framework to channel the participation of compliance entities, financial intermediaries, project developers, and more. The second critique derives from the first: if only a small handful of experts globally understand what carbon markets are and how they function, the spaces for informed debate and democratic accountability are severely diminished – we are left to take the experts at their word. This matters in obvious ways given what's at stake with global climate change, alongside the fact that many of the complexities are in fact intractable and cannot be resolved even if we get the best and brightest working on these problems.[10]

Negotiators are now working to draft the guidelines, or rulebook, for the implementation of the Paris Agreement and Article 6 in time for the 24th COP in 2018, with this work supported heavily by the global epistemic community advocating for emissions trading. Current members include the World Bank Group's Carbon Pricing Leadership Coalition, established at COP 21 in Paris, which counts amongst its members twenty-seven national and subnational governments; a who's who of the world's heavy emitters, including Barrick Gold Corporation, BP,

LafargeHolcim, Shell, and the TransCanada Corporation; heavy hitters from the global financial and investment communities; multilateral institutions and organizations, including the IMF, the OECD, and the World Economic Forum; ENGOs like EDF, The Nature Conservancy (TNC), and the World Wildlife Fund (WWF); and the International Emissions Trading Association (IETA), a global business lobby that advocates for emissions trading on behalf of its powerful membership, including the world's most influential fossil fuel, mining, and cement companies, utilities, and financial and investment firms. Working closely with the UNFCCC, the World Bank, national and subnational governments, ENGOs, and like-minded academics – typically economists – IETA enjoys unparalleled access to the world's governments and policy makers.

Before considering how market experiments are unfolding globally, one additional piece of the puzzle must be considered: specifically, what goes into making a market – the assumptions and component parts that create the whole – shapes critically what it does, or doesn't, do. In the case of carbon markets, it is the fine details of market construction that reveal a puzzle whose pieces fit poorly in place, if at all. This challenges claims that carbon markets are able to fulfil their stated purpose of effectively combatting catastrophic climate change.

Knowing what we don't know in the making of markets

As discussed above, proponents of carbon trading justify their position with reference to cost-effectiveness. In particular, neoclassical and neoliberal economic accounts of carbon trading suggest that it is the most efficient, and thus cost-effective, mitigation strategy available, relative to carbon taxation and command and control style regulation, since price discovery is said to emerge via supply and demand relative to the capacity of firms to lower emissions within a set cap. Incentivized to take advantage of efficiency gains, those firms with lower marginal abatement costs can reduce emissions more efficiently than those with higher costs, selling their unused carbon allowances to carbon-intensive firms, thereby realizing additional economic gains. In short, by allowing the market to coordinate where emissions reductions take place as an effect of competitive advantages in abatement, carbon trading is said to keep the cost of abatement lower than competing options, with its value reflected in the market price for carbon.[11] These arguments are intellectually and theoretically appealing since they promise to minimize the costs of acting on climate change, while offering new growth and accumulation opportunities – a win-win for the climate and economy.

This particular rendering of markets is one that understands them as largely neutral and value-free, and as mere technical instruments for achieving the greatest good. Anthropologist Tania Li offers us the concept of 'rendering technical', employing it to conceptualize the process by which actors, having identified a particular problem, seek to discipline it and make it legible for intervention. Quoting Nicolas Rose, she notes that rendering technical is a process for representing 'the domain to be governed as an intelligible field with specifiable limits and particular characteristics … defining boundaries, rendering that within them visible, assembling information about that which is included and devising techniques to mobilize the forces and entities thus revealed' (p. 57).[12] Put differently, 'rendering technical' orders, classifies, and compartmentalizes the otherwise complex and unruly, so that the techniques of governance and management have a simplified and disciplined field upon which to act. Climate change is one of the most complex problems to ever have confronted the global community; carbon weaves a dense interconnected web throughout contemporary biophysical, social, political, and economic life, present in the air we breathe and all that we touch. The use of carbon markets in response to climate change confronts these complexities with measurement and accounting, suggesting that if we succeed in getting the balance sheet right – who emits and how much, who reduces and how much, who sequesters and how much – the monetization and trading of carbon amongst self-interested rational actors will ultimately beat this thing. Thus technified, the method of intervention is rendered simultaneously non-political, a point underscored by Li. To illustrate why the technification and marketization of climate change pose such significant problems, market development can be considered along two broad tracks. The first considers the process of commensuration through which a fungible commodity is produced; and the second, the political economy of carbon market design.

Regarding the former, commodification is a fundamental prerequisite for market creation and operation, for it encloses and thus produces that thing that can ultimately be bought and sold by market participants – in this case, the tCO_2e. But how can we know that one tCO_2e is precisely equal to another tCO_2e – say, the tCO_2e *emitted* from a coal-fired power plant in Germany, and the tCO_2e that is *reduced* from land-based measures to sequester carbon allowing a project to earn carbon offsets, whether in Canada or in Mexico? This is a critical question since it is the equivalency between one tCO_2e and any other tCO_2e that is supposed to ensure the environmental integrity of the system once different actors begin to trade with each other. Crucially, environmental integrity is tied here to the task of aggressive and rapid emissions reductions in a dangerously warming world – we simply don't have space for uncertainty. And yet this is precisely what we get, given the deep and often intractable complexities that complicate how we know and understand carbon.

Global warming potentials

We can start with the Global Warming Potential (GWP) to illustrate the above point.[13] Under the Kyoto Protocol, six GHGs were eligible for inclusion in its carbon trading schemes – CO_2, methane (CH_4), nitrous oxide (N_2O), sulphur hexafluoride (SF_6), hydrofluorocarbons (HFCs), and perfluorocarbons (PFCs). As GHGs, Kyoto allowed their reduction to count towards the meeting of compliance targets; as gases that could be traded within Kyoto markets, they had to be quantified into fungible and tradeable units, yet each possesses vastly different properties in space and time. Specifically, each gas is more or less powerful in contributing to global warming based on its heat-trapping properties and its resident time in the atmosphere. Recall that CO_2 can remain in the atmosphere for decades to thousands of years. Not so for methane, which has an atmospheric lifetime in the range of decades, versus nitrous oxide at over 100 years, and sulphur hexafluoride at more than 3,000 years. Producing a fungible commodity to be traded in carbon markets requires making these gases equivalent, which the metric of the GWP allows for. Using CO_2 as the reference gas against which all others are measured, with CO_2 assigned a value of 1, the GWP estimates how many units of methane or nitrous oxide one gets for one unit of CO_2 once the physical and temporal differences between the gases are calculated. Put differently, the GWP for methane or nitrous oxide tells us how many units of CO_2 would be required to produce the same amount of atmospheric heating. The standard unit here is 1 tonne. In the IPCC's 2013 Fifth Assessment Report (AR5), methane was assigned a GWP of 28 over a time horizon of 100 years; nitrous oxide a GWP of 265; and HFC-23, a type of hydro-flourocarbon popular in the CDM offset market, a GWP of 12,400. This means the following over a 100-year time horizon: 1 tonne of CH_4 = 28 tonnes of CO_2; or, 1 tonne of HFC-23 = 12,400 tonnes of CO_2.

Within a carbon market, if a factory reduced 1 tonne of HFC-23, it would have 12,400 carbon allowances or offsets to trade then. Under Kyoto, buyers of HFC-23 offsets could then count them towards their own emissions reduction targets. In other words, by purchasing 12,400 offsets, a buyer could legally claim it had reduced its own carbon emissions by 12,400 tonnes. Important to understanding the significance of the GWP relative to the environmental integrity of carbon markets is the issue of uncertainty. Between the IPCC's Second Assessment Report (AR2) and its Fourth Assessment Report (AR4), the value of GWPs changed, based on the range of factors used to calculate them. This is illustrated in table 3.2.

HFC-23 saw a difference of 3,100 between 1995 and 2007 based on a 100-year time horizon. Between 2007 and

Table 3.2 GWP values in the IPCC's AR4 (2007) and Second Assessment Report (1995)

Lifetime	2007				1995			
	CO_2	CH_4	HFC-23	N_2O	CO_2	CH_4	HFC-23	N_2O
20 yrs	1	72	12,000	289	1	56	9,100	280
100 yrs	1	25	14,800	298	1	21	11,700	310
500 yrs	1	7.6	12,200	153	1	6.5	9,800	170

Source: IPCC AR4, https://www.ipcc.ch/publications_and_data/ar4/wg1/en/ch2s2-10-2.html; UNFCCC: http://unfccc.int/ghg_data/items/3825.php.

2013, HFC-23's GWP was revised from 14,800 to 12,400 over a 100-year time horizon, a difference of 2,400. As such, companies producing HFC-23 offsets using IPCC 2007 data would have received 2,400 more carbon offsets than they would using 2013 data, offsets bought and used to meet legally binding emissions reduction targets. That represents a significant and troubling discrepancy, particularly given the disproportionate role played by HFC-23 and other industrial gas offsetting projects in the CDM under Kyoto. At the time of writing, 30 per cent of issued CERs – CDM carbon offsets – have gone to HFC-23 projects, though they made up only 0.3 per cent of CDM projects. Overall, industrial gas projects accounted for 48 per cent of CERs in the CDM, or 1.7 per cent of projects. With over 1.84 billion CERs issued globally, 1.5 billion have been purchased by compliance entities in the EU ETS to meet their targets. Before industrial gas projects were banned from use in the EU ETS in 2013, many of these offsets were from HFC-23 and nitrous oxide projects in the global South. In sum, it is certain that the 1.5 billion CERs sold into the EU did not in fact represent 1.5 billion tonnes of carbon.

The authors writing on GWPs in AR5 noted that absolute certainty in GWP values is not possible given the limitations inherent to modelling the behavior of GHGs in highly complex biophysical systems. Researchers must select for specific parameters that may or may not include direct or indirect effects of the gas in question on the climate, feedback loops, a range of timeframes, regional or global metrics, and so forth. The report cites studies on the uncertainties of GWPs for CO_2 and methane, ranging from ±15% to ±26% for the former, to -30% to +40% for the latter. It adds that there is no equivalence between GWP and temperature or other climate variables, suggesting that the name itself – Global Warming Potential – is 'somewhat misleading'. Finally, it notes that, while GWPs are calculated for 20-, 100-, and 500-year time horizons, 'no scientific argument' exists for choosing one over the other, although the choice significantly impacts GWP values.[14] The international community's choice to use the 100-year time horizon as the standard for measuring GWPs thus determines how many carbon allowances or offsets – the tCO_2e – will be produced for each tonne of a specific GHG, with those numbers able to count as legally binding emissions reductions. In short, the GWP represents rendering technical in action, compartmentalizing and setting clear boundaries so that the unruly and uncertain can be tamed to meet our political goals.

The business-as-usual scenario
Further uncertainty arises when actual emissions, possible future emissions, and actual or possible emissions reductions must be measured and quantified. If the tradeable commodity is to count for anything in a carbon market, we must know exactly what it is counting for or against. In theory, jurisdictions developing cap-and-trade schemes should use business-as-usual (BAU), or baseline, emissions to determine the emissions cap, creating carbon allowances equal to the cap. To be effective, this requires precise emissions data from firms and entities with compliance obligations – also important since regulators must know how many carbon allowances a steel mill or cement factory must surrender under the scheme. The quality and credibility of the data vary across time and space; with experience, jurisdictions and market participants may get better at measuring and verifying their emissions, while the capacity for monitoring, oversight, and regulatory enforcement will vary globally from one jurisdiction to the next. Nevertheless, compiling accurate emissions inventories

entails significant uncertainties, while there is an incentive to overestimate baseline emissions on the part of regulators designing ETSs, and on the part of firms required to participate in them. If you say you are emitting more than you are actually emitting, you will get extra allowances. Extra allowances leave room for emissions to grow, allowing emitters to avoid the cost of purchasing carbon allowances if they are short, or having to reduce emissions at the source. Overall, the uncertainty of scientific quantification – can we measure with certainty the carbon-intensity of different mining operations, different fossil fuels, and different productive processes? – intersects with the political economic interests of market developers and participants to weaken the environmental integrity of carbon markets.

These issues are amplified with carbon offsetting. Within the CDM, JI, and the global voluntary carbon offset market, offset projects are diverse – the construction of hydro-electric dams, methane capture from garbage dumps, fuel switching (i.e., from coal to lower-carbon natural gas), capturing flared gas from oil wells, renewable energy projects, afforestation and reforestation projects, provisioning clean cookstoves to local households otherwise relying on local biomass, and even the construction of coal-fired power plants with technology that is said to reduce the carbon intensity of the plant. Though it appears counterintuitive, allowing coal-fired power plants to earn offsets makes sense under offsetting's technified logic.

In this case, a company in India that is planning to construct a coal-fired power plant could use less efficient, or subcritical, technologies that would release more carbon than if it used highly efficient, or supercritical and ultra-supercritical, technologies. Consequently, the company could develop a CDM proposal, arguing that with CDM money, known as carbon finance, it could afford to build the more efficient plant, lowering its lifetime emissions. It is worth noting, however, that China has phased out subcritical builds, with India following suit as a result of trade exposure and volatility in international coal markets. Yet both countries have received CDM finance for new coal plants because project developers submitted proposals claiming they would have no choice but to use subcritical technology without carbon finance. Indeed, CDM approval is contingent on whether or not project developers show that, without CDM money, there is no choice but to develop higher-emitting projects. In CDM lingo, this is known as 'additionality', with project developers having to prove that the CDM project would be 'additional' to what would otherwise happen. At least three issues are worth flagging here.

First, *Carbon Market Watch*, an EU-based NGO that monitors carbon markets globally, notes that Kyoto's offset mechanisms were designed to be 'technology-neutral', so that any technology that lowers emissions can qualify, even when used in fossil fuel projects that will lock-in fossil fuel combustion for decades. This is *unproblematic commensurability* at its worst – money that should be used to support aggressively a low-carbon transition instead gives new life to coal, the dirtiest and most carbon-intensive fossil fuel around. In turn, the localized health problems and environmental destruction resulting from coal-electricity generation do not matter, with accounting the only real concern here.[15]

Second, this illustrates how offsetting provides an incentive to overinflate baselines – even though subcritical builds are no longer the norm, claiming they are produces a new revenue stream once offsets are generated from avoided emissions. Indeed, offsets are commodities with financial value, with offset projects driven by the same imperatives as those that drive the broader economic system – growth and profitability. Embedding these imperatives at the heart of climate action has proven quite dangerous. One of the most extreme examples of this comes from the offset market for HFC-23 – a by-product of the production of HFC-22, commonly used in refrigeration. Studies showed that companies in China, India, and elsewhere increased their production of HFC-23 so they could then destroy it, earning offsets. With the possibility of earning up to 14,800 carbon offsets for the reduction of 1 tonne of HFC-23 (remember here the GWPs), this scheme proved highly lucrative for those companies involved.

Third, offsetting reveals the profound uncertainty involved in determining how much higher emissions would be without CDM carbon finance.[16] Lohmann's description of the CDM and JI as 'carved out of the future' is highly apt, pointing to a significant problem inherent to offsetting – the need to predict the future (p. 209). With no way of actually knowing the future, project developers produce 'counterfactual' scenarios. It starts with a measurement of the baseline scenario – actual emissions in the absence of the project – against which a counterfactual scenario is pitched: estimated emissions with the project. The calculations required to develop the two scenarios produce a higher-emissions scenario (baseline) vs a lower-emissions scenario (counterfactual). If the baseline produces 100,000 tCO_2e and the counterfactual 75,000 tCO_2e, subtracting the latter from the former determines how many offsets the project receives – 25,000 available for purchase by heavy emitters to count as their own emissions reductions.

Offsetting projects are plagued by the uncertainty of baseline and counterfactual scenarios. Bumpus' concept of 'hemming in' illustrates this, distinguishing 'between "more" or "less" uncooperative carbon'. Developing a baseline scenario requires knowing how much carbon is being emitted without the project. *Less* uncooperative carbon can be more easily measured, or hemmed – replacing coal with natural gas in a pre-existing power plant is easier to measure since the variables under consideration are fewer: the carbon content of coal versus natural gas, the plant's historical year-on-year electricity generation rates, and so forth. While still uncertain, hemming is somewhat easier. The same does not apply for project types and offset technologies dealing with *more* uncooperative carbon, as is the case for carbon forestry, clean cookstoves, small-scale water filtration projects, for example.

For the latter two project types, both have the potential to reduce emissions that occur when households, largely in the global South, collect and burn trees and biomass to boil water and cook food, releasing stored carbon. But how do we calculate how many trees were cut, how much vegetation was harvested, and how much of it was burned, year on year, to establish the baseline scenario? Determining these per household requires working with numerous variables, producing estimates based on information provided by resource users over specific timeframes and dependent on shifting household need and composition. These then intersect with changing ecosystemic conditions from one area or season to the next that are dependent on shifting weather patterns and natural occurrences, and that produce variable emission rates. This illustrates how more uncooperative carbon complicates further the production of offsets that accurately reflect emissions patterns. Moreover, the counterfactual scenario requires knowing how much the stoves or water filters will be used in the future, which is variable depending on the technology's durability, suitability, and adaptability to local socio-economic and cultural needs, and its ability to deliver real benefits. Indeed, the possibility of hemming to produce one-off, fully fungible carbon offsets is remote to impossible.

With carbon forestry projects, offsets are produced when trees are planted, or deforestation is avoided. The IPCC notes that, between 1970 and 2010, fossil fuel combustion accounted for approximately 78 per cent of global GHG emissions, with agriculture, deforestation, and land use changes combined making the second-largest contribution. Forestry is identified as a key mitigation sector, given the capacity of trees to sequester carbon. Carbon forestry was negligible in the CDM – the EU ETS banned CDM afforestation and reforestation offsets, given their uncertainty – with the global voluntary offset market, used primarily by companies to meet corporate social responsibility (CSR) standards, serving as the main market into which offsets from avoided deforestation – known as Reducing Emissions from Deforestation and Forest Degradation (REDD+) – have gone. The Paris Agreement sets the stage for the inclusion of REDD+ in a global carbon market, while approval by the International Civil Aviation Organization (ICAO) of its Carbon Offset Reduction Scheme for International Aviation (CORSIA), an offsetting scheme to start in 2021 that is meant to address rapidly growing emissions in that sector, may provide significant demand for REDD+ credits. Nevertheless, producing the baseline and counterfactual scenarios for carbon forestry is notoriously imprecise.

Defining a 'forest' is contested, given vast geographic differences globally, uncertainty about what constitutes canopy cover and acceptable thresholds for it, disagreement over the role of biodiversity in forest ecosystems, and sharp variations in the capacity of tree species to sequester carbon generally and over time. For instance, scientific research shows that the Amazon rainforest became carbon-neutral due to droughts in 2005 and 2010, its carbon sequestration functions halted. This complicates tremendously the assumptions that go into producing baseline and counterfactual scenarios on the capacity of trees to store carbon, with dynamic and complex ecosystems resisting the orderly technification required to produce fungible offsets. Trees are also impermanent, subject to forest fires, pest infestations, and natural disasters, while offset projects may simply displace deforestation activities outside the project area – logging, ranching, agriculture, and so forth. Quantifying historical rates of deforestation, its causes, and how socio-economic and political trends shape it, further limit the possibility of constructing a straightforward balance sheet – trees destroyed, trees planted, trees saved – to produce the fungible tCO_2e. A recent study on REDD+ offsetting in Madagascar and the Democratic Republic of the Congo to determine the environmental credibility of baseline scenarios found that their development was faulty on a number of counts, including the fact that historical deforestation rates were estimated for sites – known as reference areas – experiencing much higher rates of deforestation than what was occurring in the project areas themselves. These high rates were then used to suggest that deforestation within the project areas was higher than it actually was, allowing project developers to overinflate baseline scenarios, earning the projects more offsets. The researchers linked the tendency to overinflate baseline scenarios in part to business models and the need to generate sufficient revenue, showing that private certification proceeded despite these flaws.

Critics of land-based carbon offsetting further highlight the lack of equivalence between terrestrial and fossil carbon. Terrestrial carbon actively cycles through the planet's vegetation, oceans, and atmosphere, while fossil carbon is stored in underground pools that took many millions of years to develop. As a prisoner of time and geology, fossil carbon only participates in the active carbon cycle once it's dug up and burned. Doing so dramatically expands the active pool of carbon, with fossil fuel combustion overwhelmingly responsible for current rates of global warming. By licensing the continued combustion of fossil fuels, forestry offsets impede aggressive action on climate change. Rates of deforestation must be slowed dramatically, and for many reasons, but doing so should not sustain the high-carbon, fossil-fuel-intensive system at the heart of the problem.

These examples show how scientific uncertainty and political economic interests intersect to diminish the credibility of the tCO_2e. In each, the future factors heavily since producing the tCO_2e demands assumptions, estimations, and models of what the future will look like, meaning choices are simultaneously made over what to leave out. With these assumptions then tied to baseline scenarios – and, let us be clear, today's business-as-usual baseline is built on a present dripping with carbon-intensive fossil fuels – the scenario mirrors a particular brand of growth-driven fossil capitalism, with heavy emissions that hold constant. The result is more offsets and more tonnes of CO_2e, and more property rights allowing high emitters to keep on emitting, than would otherwise be the case in an alternative future scenario. Indeed, by rendering technical climate change and its solutions, the problem is rendered non-political. In short, offsetting and trading smooth out the threats to the current political economic order, ensuring minimal disruption to those who benefit most from it.

But what if we could be fully certain in producing the tCO_2e? If this were the case, how effective is carbon trading in supporting aggressive emissions reductions in line with the science? The answer: not very effective at all. To understand why, we turn to the political economy of carbon market design, for it shows us in no uncertain terms that markets are political constructs shaped by existing relations of power. In this, the world's existing markets have been designed so that low prices and loopholes become norms rather than anomalies.

The political economy of carbon market design

Our earlier overview of emissions trading schemes globally provided carbon prices per tCO_2e for each individual ETS. They ranged from a high of US$18.46 in South Korea to a low of $0.87 in the Chongqing Pilot ETS in China. Still lower were CDM offsets at $0.25. Taken together, the average carbon price was $7.60 per tCO_2e. While each scheme is distinct in terms of coverage and purpose, reviewing actually existing prices in carbon markets worldwide is revealing. In *Carbon Pricing Watch 2017*, the World Bank and Ecofys note that the roughly three-quarters of emissions globally covered by a carbon price, including carbon taxes and markets, are priced below $10 per tCO_2e. The 2017 *Report of the High-Level Commission on Carbon Prices* notes that to stay well below 2 °C of warming, carbon prices need to be in the range of at least $40–80/$tCO_2e$ by 2020, and $50–100/$tCO_2e$ by 2030. Others suggest that carbon prices must be even higher to meet the 2 °C target – $150 to $200 per tCO_2e, or more. And what would prices need to be to keep us below 1.5 °C of warming? Recall that the stated goal of carbon pricing is to make it *expensive* to engage in high-carbon activities, prompting the search for low-carbon alternatives. Yet current prices in existing carbon markets worldwide are abysmal by this measure. Indeed, in the case of the EU ETS, the world's largest and longest-running carbon market, market analysts with Thomson Reuters titled a September 2016 webinar 'Carbon Pricing: A Missed Opportunity?', asserting that the regional ETS, covering twenty-eight EU member nations plus Iceland, Lichtenstein, and Norway, and in place since 2005, was playing no role in the emissions reductions taking place in the bloc. The reason? Low prices.

While some carbon market proponents acknowledge that prices globally are too low, they commonly blame poor design and government interference in the market for this predicament. Economists will point to economic models that show the efficiency and effectiveness of carbon markets as abatement tools, bemoaning policy makers who continue to shackle the invisible hand and the magic it might work. Yet the act of economic modelling is one of uncertainty – it makes assumptions and value judgements, it isolates and abstracts from existing realities, it idealizes, and it selects for some variables over others. In this way, economic modelling enframes and renders technical, and, thus, power and politics in the making of markets – not to mention their operation – are largely disavowed. What economists tell us should happen never really does, precisely because the real world and the power relations that shape it influence profoundly how policy gets done, especially when that policy strikes at the heart of our current political economic system.

It is widely acknowledged that low prices in carbon markets worldwide are due to market oversupply. That is, there are too many allowances and offsets available, meaning there is no scarcity to push prices higher. Prior to the crisis of oversupply in the EU ETS and the CDM, prices looked much different. In 2008, the price of carbon allowances in the EU ETS, or European Union Allowances (EUAs), was around €30 per tCO$_2$e, while the price of CERs in the CDM was around €23 per tCO$_2$e. From there, both began significant supply-driven declines – carbon was rated the world's worst-performing commodity in 2011 – to hit near bottom. At the time of writing, EUAs were around €7 per tCO$_2$e, with CERs valued at around $0.25. Data shows that the global carbon market, including existing ETSs and Kyoto offset mechanisms, was valued at $176 billion in 2011, while in 2017 its value is approximately $32.8 billion. From 2005, the World Bank had published its annual *State and Trends of the Carbon Market*, suddenly cancelling it in 2013 and replacing it with *Mapping Carbon Pricing Initiatives*, replaced a year later with its *State and Trends of Carbon Pricing* series, co-published with Ecofys. Indeed, something was deeply amiss in carbon markets globally, something the Bank and other vocal proponents could no longer ignore.

So why are prices so low? On the one hand, the price of carbon is influenced by events and circumstances outside of the market. Economic slow-downs and crises, the increased introduction of renewable energy options, improved energy efficiency, amongst other things, lower emissions and decrease demand within an ETS. While important, however, they form only part of the story. Arguably, the political economy of carbon market design is of greater importance, since design features lay the bedrock from which the market emerges, setting the parameters for price behaviour once it gets going. As Robertson suggests, regarding the creation of markets in environmental services, price discovery is 'an expansive and time-consuming task of institutional construction', with supply and demand 'entirely created by regulatory directives rather than the utility functions of individual participants' (pp. 501, 504).[17] In short, regulators set a cap (the supply), with compliance entities demanding allowances or offsets only to the extent that the rules and regulations apply to them. Those rules and regulations thus matter profoundly. In terms of their influence on price behaviour and their contribution to market oversupply, three especially important design features are: the setting of emissions caps; the inclusion of provisions to allow for the use of carbon offsets to meet reduction targets; and the awarding of free carbon allowances to heavy emitters – although the latter presents a more complicated story in terms of how it influences prices.

Cap setting

In theory, setting an emissions cap within an ETS should be done in line with the science of climate change – so how much should we emit to stay below 2 °C or 1.5 °C? – and it should account for historical emissions so that policy makers can measure one against the other – if emissions are here (x), and we need to get there (y) to limit warming to 2 °C, where does the cap (z) need to be? It is no secret that no one is setting caps to keep us below 2 °C of warming. Instead, cap setting is highly politicized, given the economic implications of a stringent cap – if it is extremely rigorous, there will be a shortage of allowances, resulting in a very high price. As market analysts noted with respect to cap setting in the *Carbon Markets Almanac 2014*, 'The experience in the EU, and also in other ETSs like RGGI and the Californian system, shows that policy makers often *do not dare* set stringent targets which really bolster emissions reductions in the medium run – in most systems the cap is not even really binding as it is set at a very generous level' (p. 8 – emphasis added).[18] Setting caps too high above BAU emissions at the start of the ETS is known as 'overallocation', in that the market is allocated too many emissions allowances.

In Phase I (2005–7) of the EU ETS, estimates suggest the

cap was 4–5.6 per cent above BAU emissions, eventually contributing to a crash in the price of EUAs to €0.10 per tCO_2e when the depths of oversupply became apparent. In the RGGI, an electricity-sector cap-and-trade programme covering nine northeastern US states, the Phase I (2009–11) cap for annual emissions was set at 188 million short tons of CO_2, while emissions only came in at 124 million tons in 2009, 138 million in 2010, and 124 tons in 2011. The cap was then reduced to 165 million tons of CO_2, still well above emissions during Phase I, with emissions in 2013 only reaching 86.5 million tons. A further revision to 91 million tons in 2014 still failed to reflect actual emissions – they came in that year at 86 million tons of CO_2. In 2011, RGGI allowance prices were at the floor price – a limit set by regulators below which it could not go – of $1.89. While coal-to-gas switching in the RGGI states, and the fallout from the subprime mortgage crisis in the USA, impacted emissions, policy makers consistently failed to align the cap with actual emissions. Estimates suggest California's 2013 cap was roughly 12 per cent higher than emissions in 2011 from covered entities; in Quebec, observers suggest that the market was allocated above the cap, meaning too many allowances were created, increasing the cap by virtue of extra supply; and China's seven pilot ETSs have been described as massively overallocated by millions of allowances, with studies suggesting that economic assessments to determine BAU emissions prior to introduction in a number of pilots were absent.[19] While this discussion is brief, the point to underscore is that, when caps are set too high, oversupply is cooked into the cake, so to speak, and compounded then by additional design features.

Offsetting

By providing compliance entities with an additional source of carbon credits that tend to sell at a discount relative to government-issued allowances, offsets inflate caps beyond their initial levels by injecting a new source of supply into the market. Lohmann describes it well when he states that offsetting allows governments to claim that they are tightening caps while simultaneously punching holes in them through which outside supply flows in.[20] Worldwide ETSs display different rules on the use of offsets. In the EU ETS, compliance entities could meet up to 50 per cent of their reductions during Phase II (2008–12) and Phase III (2013–20) with offsets from Kyoto's JI or the CDM; and New Zealand's ETS, started in 2008, allowed compliance entities to meet 100 per cent of their obligations with Kyoto offsets until 2015 when the government's decision not to join a second Kyoto commitment period barred the nation from using Kyoto offsets. The linked markets of California, Quebec, and Ontario allow emitters to meet up to 8 per cent of their reductions with offsets, which, in the case of California, must be produced in the USA, while Quebec accepts offsets sourced provincially or from the USA, and Ontario accepts offsets produced anywhere in Canada. South Korea's ETS allows compliance entities to meet up to 10 per cent of their obligations with domestically sourced offsets, while offset rules in China's pilot ETSs vary.

The experiences of the EU ETS and the CDM show how offsets, by inflating caps, contribute to oversupply in carbon markets, dragging prices down. By 2013, the EU ETS was heavily oversupplied by approximately 2 billion excess allowances, of which some 455 million were offsets. The 50 per cent limit on offset use from 2008 to 2020 allows European emitters to use up to 1.6 billion Kyoto offsets during that time. By 2015, they had used 1.43 billion – 866 million from the CDM and 570 million from JI. The impact of offsets on EUAs, by increasing supply when oversupply was already a problem, contributed to crashing prices. Moreover, with offsets trading at a discount to carbon allowances, the price of the former necessarily pulled down the latter. Using the CDM as an example, during Phase II of the EU ETS its projects received 1.48 billion CERs, with another 168 million awarded by early February 2016, totalling roughly 1.65 billion. This is well beyond what the EU could absorb; when it became apparent that too many CDM projects were being registered and awarded CERs, prices began to crash. Worth highlighting, too, is that the CDM was driven by its own set of dynamics – when prices were high, projects in the global South expanded rapidly as developers and investors sought to profit from this emerging global market. Yet the CDM's own crisis of oversupply, combined with dwindling demand from the EU and the passing of EU legislation barring CERs from industrial gas projects and limiting them to CERs produced in the world's LDCs, has meant that the CDM is barely surviving. New Zealand offers another stark example – when covered emitters could meet 100 per cent of their obligations with Kyoto offsets, the market was flooded. When Kyoto offset prices plunged, New Zealand carbon allowances followed suit, sinking to under $2 per tCO_2e in 2013. While newer markets have learned from some of these mistakes, implementing lower thresholds for offset use, the fact remains that they are intended to punch holes in caps already insufficient to meet a 2 °C target, let alone 1.5 °C. In short, they work to contain costs that are already far too low.[21]

Free allowances
The practice of giving free allowances to emitters in ETSs undermines their effectiveness in a number of ways. Heavy

emitters and governments argue for free allocation given the possibility of carbon leakage. Leakage is said to occur when producers, especially energy-intensive and trade-exposed producers, are placed at a competitive disadvantage relative to producers in jurisdictions without a price on carbon, causing them to relocate outside of the jurisdiction introducing an ETS. If this happens, emissions simply shift elsewhere. While the evidence on leakage is not particularly robust, emitters lobby heavily for free allowances, with many arguing that buy-in from carbon-intensive firms is required to get policy support. In the EU ETS, 95% and 90% of allowances were freely allocated during Phases I and II respectively; Phase III required utilities to purchase allowances, giving 97% of industrial emitters 100% of their allowances for free. California's Cap-and-Trade Program freely allocated 100% of allowances to entities deemed at risk during Phase I, introducing a sliding risk scale for Phases II (2015–17) and III (2018–20). Those in the high-risk category continue to receive 100% free allowances; those in the medium-risk category receive 75 and 50% of their allowances for free during Phases II and III, respectively; and those at low risk receive free allowances at 50% and 30% during the two phases. South Korea freely allocated 100% of allowances in Phase I (2015–17), and will allocate 97% in Phase II (2018–20), and less than 90% in Phase III (2021–5). In short, free allocation is common to all ETSs, although problematic on a number of fronts.

Importantly, it shields carbon-intensive firms and sectors – precisely those who need to lower their emissions rapidly – from carbon prices. In this, the price signal, which is supposed to change high-carbon behaviours, cannot reach those whose behaviour it is supposed to change. Consequently, there is little to no demand from those receiving free allowances, something that might otherwise put upward pressure on prices. Equally problematic, during Phase II in the EU ETS, emissions from industrial entities were roughly 3.42 billion tCO_2e, yet they received 4.42 billion free allowances – 1 billion more than needed. This meant that some firms, including ArcelorMittal, the world's largest steel company, possessed roughly 140 million surplus allowances. When benchmarking was introduced in the EU ETS for Phase III, using industry efficiency standards instead of grandfathering to determine free allocation, heavy emitters continued to receive surplus free allowances – over 2013–14, the EU industrial sector emitted 1.48 billion tCO_2e, receiving 1.52 billion free allowances. Subsequently, many of these surplus free allowances have entered the market. As commodities, carbon allowances become company assets for those firms holding them. Studies show that, between 2008 and 2014, carbon-intensive industrials in nineteen countries in the EU ETS generated 'windfall profits' of €8.1 billion by selling surplus free allowances.[22] Selling carbon assets helps firms to raise revenue when required, while simultaneously putting downward pressure on prices in an oversupplied market – hardly an effective and urgent strategy to tackle dangerous climate change, it would seem.

In summary, as a result of the specific political economy of market design, we see ETS regulations that, to varying degrees, engender market oversupply and low prices. In ETSs that have price floors, including those in California and Quebec, prices hover around the floor; while their prices are higher, it is only because governments legislated a floor below which they cannot go. This highlights the availability of tools for managing price volatility and market operations. In addition to its price floor that increases 5 per cent per year plus inflation, California restricts participation in its auctions so that non-covered entities can purchase no more than 4 per cent of available allowances. This contrasts with the absence of such restrictions in RGGI where speculators hold up to 80 per cent of allowances, with significant power to influence market behaviour according to imperatives that have nothing to do with addressing climate change. In the EU, too, speculators play a large role in the ETS. California also limits how many allowances entities can possess at any given time, to deter market manipulation. Overall, however, higher prices in places like California, Quebec, and Ontario are nothing more than high low prices.

Returning to offsets, two recent reports commissioned by the European Commission provide damning assessments of the environmental integrity of Kyoto's flexible mechanisms. The first examined the CDM to determine how additional it was. Recall that approval requires projects to be additional to business as usual, bringing us back to baselines. In their exhaustive study, the authors concluded, based on a statistically significant sampling of CDM projects, that a stunning 85% of analysed projects, and 73% of CER supply that could be delivered through to 2020, 'have a low likelihood that emission reductions are additional and are not over-estimated'.[23] Findings suggest only 2% of the projects analysed and 7% of potential CERs to 2020 have a 'high likelihood of ensuring that emissions reductions are additional and are not over-estimated'. Their overall conclusion: 'the CDM still has fundamental flaws in terms of overall environmental integrity' (p. 11). Consequently, they caution heavily against relying on offsetting in the development of a global carbon market under the Paris Agreement, with implications for the emergence of ICAO's CORSIA in 2021. In the second report, a review of JI found that, of the sixty projects analysed, 73% issued carbon offsets that were 'unlikely to be additional', while 80% of project types produced offsets that had 'questionable or low environmental integrity'. The analysts suggest that 'JI may have enabled global GHG emissions to be about 600 million tCO_2e higher than they would have otherwise been' (Kollmus et al., p. 5), and that auditors did not fulfil their responsibilities to ensure the environmental integrity of the projects.[24]

Both studies reveal a crucial flaw in carbon offsetting. Specifically, the authors of the JI study note that offsetting is about 'keeping global emissions the same' (p. 5). Licensing higher emissions on the part of the emitter, the offset is then designed only to cancel them out; it is not designed to produce emissions reductions. Yet reductions are precisely what we need in the face of catastrophic climate change. Moreover, when offsets themselves aren't additional and lack environmental integrity, rather than staying the same, emissions continue to increase. And thus we see how 'flexibility' through offsetting is akin to Newspeak in Orwell's fictional *Nineteen Eighty-Four*, shackling our range of thought so that business as usual dons the robes of meaningful action. And so we are back where we started, yet time is not on our side.

Conclusion

Uncertainty, low prices, overinflated baselines, gaming the market, substandard auditing, unrealistic assumptions: is this how we should be describing climate action at this stage of the game? While some projects and trades take place that are not beset by each of these particular problems, as a whole, carbon markets represent a project to render technical climate change and its solutions, and to reduce carbon and the socio-political and economic relations that shape it to technical questions of accounting and quantification. This chapter has traced the process through which technification proceeds so that the reader is better positioned to question these markets, particularly when other options exist. If a price is to be put on carbon, why not tax it, with policies to ensure that the vulnerable and marginalized are not burdened by a cost that should be paid by high emitters? Why not tax it at levels sufficient to generate meaningful revenue and use that revenue to fund aggressively zero-carbon transitions that might benefit society as a whole while paying climate debts? Better yet, why not legislate required emissions reductions as has been done successfully in many parts of the world, not leaving mitigation dependent on politically structured market conditions that crash prices? Rock-bottom prices not only provide little incentive to undertake emissions reductions, but also leave investments in the market subject to one's ability to turn a profit. When carbon prices globally began to crash in 2010 and 2011, investors began to flee from carbon markets because the return was insufficient to keep them there. This begs the question: what kind of mitigation strategy is this? And here we cut to the core, since carbon markets prioritize the profit motive first – at what point are my profits threatened by a carbon price so that I must act? Is there an opportunity to profit from investing in offset projects, and at what point is the return on investment insufficient to keep me in the game? Can I buy low today and sell high tomorrow to turn a profit for my clients? How much carbon can I buy so that I have sufficient power to influence the price? Do I care that, when I offload my carbon, I may sink the price, a price that is supposed to save us from catastrophic climate change?

The next chapter will begin to explore how the problem of carbon is being addressed outside of carbon markets, strictly speaking, so that we might consider some of the other options on offer, good and bad. Carbon markets are not the only game in town, although many are working hard so that they might be. This demands that we consider those alternative futures that markets, and the ceaseless drive to commodify ever more of the socio-natural world, make otherwise invisible. And this matters deeply to the question of democracy, a topic whose exploration will close this book. Carbon markets insulate privilege; they allow those with the ability to pay to trade and offset so that they might continue with their carbon-intensive production and consumption. They prolong a business-as-usual cycle dripping with carbon-intensive fossil fuels and within which their privilege insulates them to a much greater degree from the effects of climate change. These markets are thus classed, racialized, and gendered in profound ways, betraying a spatialization of responsibility that has little to do with the history of actual carbon emissions the world over, emissions that have generated untold wealth alongside untold poverty. Climate change mitigation is not a technical matter, and it cuts to the core of who we are and who we might be. It is time to treat it as such.

CHAPTER FOUR

Carbon Transitions

The problem of too much carbon is unparallelled in terms of its urgency and complexity. Skim the pages of chapter 1 in this book for a recap of where catastrophic climate change is taking us – deadly heat waves, drought, flooding, devastating storms, collapsing food systems, mass extinction, biodiversity loss, the increased spread of infectious diseases, sea-level rise wiping out cities and communities, climate refugees, misery, and suffering. This morning's headlines suggest that, by 2100, 'deadly heat may threaten the majority of humankind', and that a 'climate emergency' is underway as 'fish abandon tropical waters'. It is necessarily the poorest and most vulnerable who suffer the worst of these effects – air conditioning a luxury for those who can afford it, the livelihoods of fishing communities in the global South destroyed as their sources of bodily sustenance and income seek out cooler waters. Data from a recent report, 'Banking on Climate Change: Fossil Fuel Finance Report Card 2017', shows the world's largest banks financing the most carbon-intensive extreme fossil fuels – coal, liquefied natural gas, and tar sands, Arctic, and ultra-deepwater oil – to the tune of $290 billion between 2014 and 2016, suggesting that the gulf between acknowledging cause and acknowledging effect remains stunningly wide.

Harnessing hope against this backdrop is no easy task; and yet, without hope, one wonders what outcomes might flow from an approach that gives us too little to hold on to in the struggle for a better future. In his book *Violence*, Slavoj Žižek writes of what he calls a 'fake sense of urgency' pervading left-liberal humanitarianism. He sees it in statements like the following: 'In the time it takes you to read this paragraph, ten children will die of hunger' – or when Starbucks papers its store entrances with posters telling of profits from coffee sales providing healthcare to poor Guatemalan children. It is the 'urgent injunction' embedded within each, the call to immediate action, that proves dangerous for Žižek, for they imply the time to think has passed, replaced instead by the need to 'act now'. But wait – do we dare ask why Starbucks doesn't pay coffee farmers living wages in the first place? Would that not be a better place to start, instead of hoping that the company might voluntarily return a tiny fraction of their substantial profits back to the poor, profits made possible on the backs of producers? For left-liberal humanitarians, the answer is no. Children are dying now, they need healthcare now, and we should be grateful that Starbucks is willing to do something it doesn't have to do. It is a message premised on the idea that there is no time to think, no time to learn, only time to act. And in this case, it is the act that then fortifies simultaneously those structures that produce the need for the act in the first place, since we fail to call them into question, thereby normalizing them further. Žižek counters this with a call for 'patient, critical analysis' so that we might learn deeply about causes (pp. 6–8).[1]

One of the problems with climate change – and there are many – is that the urgency is quite real. Within the space of this urgency, people want and need answers now. Does the fact that climate change is a different kind of beast, its noose tightening with each passing day, therefore mean that the space for patient, critical analysis disappeared long ago – maybe we had it in 1987 with the publication of *Our Common Future*, or in 1992 with the Rio Earth Summit and the adoption of the UNFCCC, or maybe even with the adoption of the Kyoto Protocol in 1997, but surely now it is gone? The argument here is that now, more than ever, we need patient, critical analysis. Climate change is a problem that defies simplification; if one does not truly understand it – its origins, causes, and impacts – then one is ill prepared to really and truly do something about it, as we see with responses like carbon trading, a strategy that delays the kinds of aggressive policies and interventions that correspond to our current crisis – over 1 °C of warming with no end in sight. In fact, heeding the urgent injunction to 'act now' makes us complicit, for it demands that we normalize those structures and relations of power that exist dialectically with a warming world, each sustaining the other, allowing these same structures and relations of power to set the parameters within which action will take place. If such is the case, actions will be mere Band-aids, under which the wound continues to fester. And so we are confronted with a challenge: if we are honest about the origins and causes of climate change, then we admit the need for deep structural change. Admitting this need may prove so overwhelming, however, that there is little reason for hope. So how do we proceed?

The concept of pathways may be useful in this respect.[2] Pathways have a point of origin, boundaries, and destinations. We are thus able to deduce where they came from, and where they will take us. At the same time, they may crisscross and overlap, and they are dynamic, their boundaries shifting with time. The decision to choose which pathway to take has much to do with where one wants to go and how one wants to get there, with some much better placed to argue in favour of their pathway, ensuring its grooves and surface receive much higher traffic, for better or worse. In thinking through the myriad strategies, policies, projects, and approaches that exist to address climate change, we can thus place them on their respective pathways. In so doing, we are better situated to consider from where and whom they came, where they propose to take us, and how they plan to do it. Equally as important, invoking the concept of the pathway to analyse or propose action on climate change can be an important counter to paralysis. Pathways are, by their nature, processual; there is a process that gets one from A to B, and to C and beyond. It may be slow and it may be daunting, but if we conceptualize change as a process, taking the time to analyse the range of actors, institutions, and interests involved, and the implications of each, we are better positioned to choose the right path, while calling out those that offer false hope and false promises.

Indeed, patient, critical analysis requires that we connect the dots when considering proposed solutions to the problem of carbon – who benefits and who pays, what is enabled and what might be blocked. In dealing with transportation emissions, for example, support for electric vehicles (EVs) has reached fever pitch, with governments and companies the world over promising vast new fleets of EVs to satisfy modern society's thirst for automobility. In many ways, the promise of a world made mobile by ever more EVs blunts concerns about the damaging role of the personal motor vehicle in modern society. The production of EVs remains highly resource-intensive, with research suggesting that the mining of rare earth minerals for battery production, while environmentally and socially problematic in some regions of the world, will be difficult to sustain at levels required to meet climate change targets. Policies that encourage electrified personal automobility often run counter to those focused on rethinking how we organize our cities and neighbourhoods, investing instead in liveable, walkable communities, connected by highly efficient and affordable public transportation networks. Such policies discourage us from thinking too seriously about the devastating role played by the car and car culture in critical habitat loss and ecosystem destruction. Rebates encouraging the purchase of EVs tend to benefit upper-middle-class and elite consumers able to afford these vehicles, while public transportation, more critical to the daily lives of poorer sectors of society, remains underfunded. In short, if the only relevant question is whether or not the car is electric, we fail to connect the dots. Competing strategies to deal with carbon will entail competition and trade-offs. Large auto manufacturers will harness their significant class power to ensure that the narrative of a green future is one that keeps us driving and consuming, with a minor tweak. The argument here is that it is dangerous to isolate carbon so that patient, critical analysis is not part of the process.

So what does this mean in concrete terms? If one types 'low-carbon transition' into an internet search engine, approximately 4.1 million hits come up; 8.3 million for 'low carbon economy'. Indeed, the buzz is loud when it comes to talk of transitions, but what are we to make of our options? Carbon capture and storage (CCS), solar radiation management, natural gas, biofuels, wind, solar, geothermal energy – each of these, and many more, travel particular pathways depending on their origins, boundaries, and destinations. While in some cases the technology may be the same – take solar, for example – how it is owned and accessed may result in very different pathways, and very different outcomes. Rather than providing an exhaustive overview of all possible pathways, this chapter will start with those that we want to avoid, moving progressively to those that offer options and alternatives for where we might want to go in a carbon-constrained world. While not entirely prescriptive, it argues that those pathways worth pursuing are so because, in addition to offering real strategies for effective emissions reductions, they are strategies embedded simultaneously in a larger project of social transformation. Social transformation here implies social, political, economic, and environmental justice – luxuries reserved for the few under the current world order of fossil capitalism.

In this, low- and zero-carbon pathways should not restrict the ability and capacity of citizens to access those resources necessary for survival; they should not actively delay a transition away from fossil fuels, locking in carbon-intensive infrastructure for decades to come; they should not reduce nature to a commodity, necessarily privileging those who can afford to pay; they should not hasten a process through which citizens are increasingly reduced to consumers; and they should not further restrict the spaces within which democratic decision-making might flourish. These are difficult demands at this particular moment, defined as it is by advanced neoliberal globalization, growing global inequality, and the continued concentration of power in elite spaces, but they are possible, and there are examples from which we can draw. It is to this task that we now turn.

The yellow brick road

In L. Frank Baum's fictional work *The Wonderful Wizard of Oz*, the main character Dorothy and her companions must follow the golden yellow brick road to the Emerald City to find the Wizard, for only he can help Dorothy get back to home and hearth in Kansas. When it is finally revealed that the Wizard is no Wizard at all, but a mere simple man who, by accident, found himself in the strange land of Oz, whose people feared him because he had descended from the clouds in his hot-air balloon, we also learn the truth of the Emerald City. The exchange between the Wizard and Dorothy proceeds thus:

> 'Just to amuse myself, and keep the good people busy, I ordered them to build this City, and my Palace; and they did it all willingly and well. Then I thought, as the country was so green and beautiful, I would call it the Emerald City; and to make the name fit better I put green spectacles on all the people, so that everything they saw was green.'
> 'But isn't everything here green?' asked Dorothy.
> 'No more than in any other city', replied Oz; 'but when you wear green spectacles, why of course everything you see looks green to you.'[3] (p. 79)

This exchange is a fitting place to start, for some of the pathways on offer today require a journey down the yellow brick road, green-coloured glasses secured in place so that we might view the Emerald City – a zero-carbon future – as we are told we should, but not as it is, or as it will be. Negative emissions technologies, solar radiation management, biofuels, and natural gas are just some of the pathways worth exploring in this context, for reasons to be explained below.

Starting with carbon capture and negative emissions technologies (NETs), these entail, in basic terms, capturing carbon before it is released to the atmosphere, or using technologies that support its removal once it's there. They include carbon capture and storage (CCS) for the former, and bioenergy with carbon capture and storage (BECCS), direct air capture, enhanced weathering, afforestation and reforestation, and technologies to increase carbon uptake in oceans and soils, amongst others, for the latter. Both CCS and NETs are reminiscent of Promethean approaches to environmental management. In Greek mythology, it was Prometheus who stole fire from Zeus, bringing it to humans so that their capacity to control and manipulate their environment was enhanced dramatically. It is thus that Promethean environmentalism displays a certain techno-optimism, believing that human ingenuity will unleash technological advances capable of addressing our most difficult challenges. On the other hand, many, while troubled by these technologies, support them nonetheless, believing they are the best we can hope for in a world in which political economic rivalry continues to shape state relations, and in which national commitments to the Paris Agreement, as they currently stand, put us well past 2 °C of warming above pre-industrial levels. Indeed, a certain intuitive logic can be found in this position. As the US government under Donald Trump withdraws from the Paris Agreement and as fossil fuel infrastructure continues to expand globally, the temptation to embrace CCS and NETs as options of last resort is deeply seductive. But the challenges and risks are great, requiring a closer investigation.

Carbon capture and storage
CCS has gained considerable visibility in recent years. As the name suggests, CCS entails two main processes – those dealing with the capture of carbon from power generation and/or industrial processes, and those related to the storage of that carbon once captured, although in some cases it may be used for other processes. Shell Canada's Quest CCS Project in Alberta, Canada, is an example of CCS within which the CO_2 that is produced when tar sands bitumen oil is upgraded using hydrogen in preparation for refining is captured. The captured CO_2 is then piped north and injected deep underground into geological formations. Canada's only other CCS project is the Boundary Dam CCS Project, operating in the province of Saskatchewan at the

coal-fired Boundary Dam Power Station. The CO_2 produced during power generation is captured, to be sold for use in enhanced oil recovery (EOR) in the province's Weyburn oil field. Brazil's offshore Petrobras Santos Basin Pre-Salt Oil Field CCS Project captures CO_2 from natural gas processing, injecting it into the Lula and Sapinhoá oil fields for use in EOR. Abu Dhabi's CCS Project captures waste CO_2 from steel production, piping it to oilfields for use in EOR. In the case of the latter three projects, when CO_2 is injected into oil wells, it mixes with oil that is otherwise difficult to recover – oil trapped in a rock formation, for example – increasing the miscibility of the oil so that it can be released and pumped. When the Global CCS Institute released its 2016 Global Status Report, it documented fifteen large-scale operational CCS projects globally, with a capacity to capture somewhere in the range of 35 million tonnes of CO_2 per year. By contrast, global CO_2 emissions in 2016 were close to 40 billion tonnes, meaning CCS, at present, is but a minuscule drop in the bucket. Proponents predict, however, that the technology will continue to advance, and adoption will spread globally in the coming decades.[4]

If the technology exists, and projects are proven, then why not back CCS enthusiastically? As the old saying goes, if it sounds too good to be true, it probably is. The Global CCS Institute notes that, according to the IEA, achieving a 2 °C scenario requires the capture of 4 billion tonnes of CO_2 per year by 2040. That number increases when the Paris target of 'well below' 2 °C, or its aspirational target of 1.5 °C, is the mitigation benchmark. In turn, mitigation scenarios are themselves derived from computer modelling that attempts to chart the likelihood of a specific level of warming (i.e., 2 °C, 2.5 °C, etc.) under different scenarios – business-as-usual energy use, the rapid deployment of renewables technologies, the heavy use of CCS, and so forth. As one group of researchers documented in a 2017 publication in *Nature Climate Change*, many of the 2 °C scenario models used by the IPCC for its Fifth Assessment Report (AR5) assume significant CCS deployment, noting that as many as 4,000 CCS facilities would need to be operational by 2040.[5] Yet others, writing in the same journal, have highlighted a number of major problems confronting this kind of large-scale deployment. First, noting that CCS requires sufficient capacity to store carbon globally, these researchers document the fact that proven permanent storage, commensurate with the extraction and burning of existing reserves of coal, oil, unconventional fossil fuels (tar sands, fracked natural gas, etc.), and biomass, remains 'unproven in sufficiency'. In other words, the global availability of secure places to store captured carbon, largely in geological formations deep underground, is not yet proven in sufficiency. The authors note that, while sufficient permanent storage may appear technically feasible, the level of research and investment required to prove or disprove sufficiency remains largely inadequate compared to the mammoth amounts of investment that continue to go into fossil fuel exploration and extraction. They conclude, thus, that CCS on the scale required is 'practically unrealistic', referencing the current 'market-driven setting' as a significant impediment to advancing CCS on the scale that the modelling suggests is required. Specifically, they note that, since the 1990s, governments and industries have been indicating their intention to develop CCS, with no corresponding 'relevant impact', evident in the fact that only fifteen large-scale CCS projects were operational by the end of 2016. One of the reasons for this is the exceptional cost of CCS, alongside the absence of global policy capable of incentivizing investment.[6] Hopes that the revenue raised from the sale of carbon allowances by the European

Commission as part of its Emissions Trading System could be used to spur investment and innovation in CCS have been persistently dashed due to rock-bottom carbon prices, while low carbon prices in turn provide little to no incentive to industrial and business actors to get into the CCS game. In short, almost no one is willing to foot the bill, hoping that market conditions will change in the future so that CCS becomes an attractive investment.

In the case of the Boundary Dam CCS Project that came online in 2014, leaked confidential documents revealed that the project was seriously underperforming on a number of fronts. Not only had the $1.1 billion project run into serious cost overruns, but the documents showed that in year 1 it did not operate at all, and then often only at 45 per cent capacity once it was up and running. It turned out that the technology used to capture carbon, a process owned by Shell that uses the chemical amine to capture the carbon, did not filter out all of the coal ash particles – recall that Boundary Dam is a coal-fired power plant – meaning that the amine was contaminated and performing at sub-optimal levels. SaskPower, the plant's operator, had signed a ten-year contract with the Canadian oil company Cernovus Energy, to sell the captured carbon so that Cernovus could use it in EOR. Unable to deliver the full amount of carbon, SaskPower had to pay CAD$7 million in penalties from the $9 million it had received in payment. Moreover, the CCS system itself consumes massive amounts of energy. Boundary Dam has a power-generating capacity of 150 megawatts, with the CCS system requiring 30 megawatts to run, and an additional 15–16 megawatts for CO_2 compression. That amounts to roughly 30 per cent of the plant's electricity going to run its CCS system. The significant efficiency and financial losses associated with operating the CCS plant led one expert observer to contend that other utilities would not follow suit for precisely this reason.[7]

CCS has suffered further significant setbacks, with two major announcements coming in the same week in July 2017. In the United States, the utility company Southern Co. out of Mississippi announced it was cancelling its Kemper CCS Project. The project, touted as a 'clean coal' project that would have gasified the coal used to generate electricity, capturing up to 65 per cent of the subsequent CO_2, was supposed to cost $1.8 billion based on initial estimates. Ten years later, project costs had soared to $7.5 billion, while the technology required to operate the gasifiers in conjunction with other project components was struggling. This came one month after Trump's federal budget proposal that recommended a 77 per cent cut to CCS research funding for the 2018 fiscal year. While many in the coal industry, including those that have been advocating for 'clean coal', were shocked, given Trump's outspoken support for the industry, the budget proposal nevertheless fits an administration that denies climate change: why fund CCS if climate change isn't really a problem? That same week in the Netherlands, utilities Engie and Uniper announced that they would be pulling out of the ROAD pilot CCS project that would have captured CO_2 from Uniper's Maasvlakte coal-fired power plant, piping it for storage under the North Sea. Observers note that extremely low carbon prices in the EU have allowed for widespread coal-fired power generation to continue with little cost, meaning that governments and industry have little financial incentive to invest in CCS. The Netherlands government intends to explore possible options to keep the project going, which at this point is the EU's only remaining CCS project, with others abandoned due to high costs.

Negative emissions technologies

With respect to NETs, the modelling of scenarios under which we limit warming to 2 °C more or less assumes their large-scale uptake, while failing to account for the political economic and geophysical limitations blocking the way.[8] In general, the Integrated Assessment Models (IAMs) used by the IPCC to model the scenarios that give us a greater than 66 per cent chance of limiting warming to 2 °C assume a 'massive deployment of negative emissions technologies' (Anderson and Peters, p. 182). In one study, researchers noted that the IPCC's AR5 database included 116 scenarios in line with a greater than 66 per cent probability of staying below 2 °C of warming, 87 per cent of which assume the use of NETs; another noted that the IPCC in general had analysed 30 IAMs containing 900 mitigation scenarios. Of the 900, 76 contained data sufficient to estimate a global carbon budget – the quantitative estimate of how much more CO_2 can be emitted before a particular temperature target would be breached – to give us a greater than 66 per cent chance of limiting warming to 2 °C. All assume the significant utilization of NETs to achieve this goal. NETs here include BECCS, afforestation and reforestation (AR), direct air capture (DAC), enhanced weathering (EW), and technologies to increase carbon uptake in oceans and soils, amongst others. BECCS entails the use of biomass – whether forestry or agricultural residues, or crops grown specifically for this purpose – to produce energy, with subsequent CO_2 emissions captured using CCS technologies. AR relies on the planting of trees in order to expand overall global tree/forest stocks, with the trees themselves capturing atmospheric CO_2. The idea here is to increase globally the availability of natural carbon stores that allow for increased CO_2 sequestration. DAC proposes to use specific chemicals – for example, amine – to capture CO_2 from ambient air, the CO_2 then being stored deep underground in geological formations. EW relies on speeding up the natural weathering of rocks since, in that process, CO_2 is captured by dissolved minerals, forming carbonates that often end up locked away on ocean floors.

In 2015, Smith et al.'s study on the 'biophysical and economic limits' of NETs was published in *Nature Climate Change*, focusing specifically on BECCS, AR, DAC, and EW. They assessed on a per unit basis – so, per tonne of CO_2 or CO_2 equivalent – the impact of NETs when land, water, nutrient, and energy requirements were accounted for, while also considering the economic costs of implementation. Some of the most notable findings deal with land, water, and energy requirements, and investment needs. For BECCS and AR, land requirements globally are significant, estimated in 2100 to be between 380 and 700 million hectares to remove 3.3 $GtCO_2e$ per year for BECCS, and around 320 and 970 million hectares to remove 1.1 and 3.3 $GtCO_2$ per year respectively for AR. To put this into perspective, the area of Canada, the second-largest country in the world, is approximately 998.5 million hectares; Russia, the largest country in the world, comes in at 1.71 billion hectares. Land requirements for EW are also large – 2 to 10 million hectares per year in 2100 to remove 0.2 or 1.0 $GtCO_2e$, respectively – while they are very low for DAC. As living technologies, so to speak, the water requirements for BECCS and AR are also significant: 720 cubic kilometres (km^3) per year in 2100 to remove 3.3 $GtCO_2e$ for the former, and 370 km^3 or 1,040 km^3 per year to capture 1.1 or 3.3 $GtCO_2e$ respectively for the latter, annually. Water needs may also be high for DAC, requiring between 10 and 300 km^3 per year to remove 3.3 $GtCO_2$ on an annual basis, while they are low for EW, as are its land requirements.

Energy requirements for AR are quite low, while BECCS

can serve as an energy source. However, the energy needs of EW and DAC are very high. The authors of the study note that, for DAC using amine, removing 3.3 $GtCO_2e$ per year by 2100 would require 156 exajoules per year when total energy costs are factored in. For a sense of the magnitude of this figure, total global energy use in 2013 was 540 EJ, so 156 EJ is equivalent to 29 per cent of total global energy use in that year. EW, to remove 0.2–1.0 $GtCO_2e$ per year by 2100, would require up to 46 EJ per year. This leads the authors to conclude that the energy needs of these two NETs could pose a 'major limitation' on their adoption without a change in circumstances.

As for levels of investment required to support the uptake of these technologies, DAC remains prohibitive, with estimates suggesting that DAC technologies, from capture, to transportation, through to storage, cost somewhere in the range of \$1,600–\$2,080 per tCO_2e. EW costs are estimated to be between \$88 and \$2,120 per tCO_2e, with a mean of \$1,104; while the mean cost for BECCS in 2100 is estimated to be \$132 per tCO_2 – this derived from six different IAMs – and between \$65 and \$108 for AR. Viewed differently, the authors suggest that investment per year by 2050 in BECCS to produce electricity, or in biofuels, must be \$138 billion or \$123 billion, respectively. They note that the investment needs are higher for EW, and even higher for DAC, while they are lower for AR.

So what are we to take from all of this? What Smith and his co-authors show quite convincingly is that NETs exist not in isolation but relative to each other, and in relation to other external but related factors. Imagine the amount of global land and water required to implement bioenergy with CCS, afforestation and reforestation, and enhanced weathering simultaneously. The global rush to produce biofuel crops in the latter half of the first decade of the

2000s is blamed in part for the ensuing global food crisis of 2007/8 that diverted food crops to energy crops, leading to a spike in the price of staple food crops around the world. This hit the poorest and most vulnerable with swift severity as the cost of food drifted further and further out of reach, inspiring food riots in cities around the globe. As the problems inherent to diverting land used to grow food to land for energy crops became increasingly apparent, many argued in favour of second-generation biofuels – non-food crops – yet failed to consider seriously the fact that non-food crops still required significant amounts of land. This led to a new round of arguments in favour of using only 'marginal' lands, yet this argument failed to appreciate that, within the current global context, 'marginal' lands are in many cases lands that are used by the poor, or marginalized groups lacking formal title to their lands (women and indigenous populations, for example) who are thus invisible to policy makers, and that these groups are pushed off their land to make way for biofuel crops. Many have studied these phenomena as part of the global trend of land-grabbing, or green-grabbing, whereby land is acquired by private investors, governments, or individuals, often forcing local citizens off their land, predominantly in the global South.[9]

There are important lessons to be learned from these experiences globally. As the need for, and/or competition over, land and water continue to grow globally, for agriculture and other forms of human use, and as climate change threatens the agricultural viability of lands and the availability of fresh water sources due to heat waves, drought, flooding, and sea-level rise, we should exercise great caution in pursuing strategies that ignore how NETs intersect with other socio-ecological variables. If we add economic cost to our analysis, the lessons from CCS are instructive – the economic incentive to invest in these technologies remains wholly insufficient when measured against the urgency of lowering CO_2 emissions globally. In closing their article, Smith et al. provide an important warning:

> Although some NETs could offer added environmental benefits (for example, improved soil carbon storage), a heavy reliance on NETs in the future, if used as a means to allow continued use of fossil fuels in the present, is extremely risky, as our ability to stabilize the climate at <2 °C declines as cumulative emissions increase. A failure of NETs to deliver expected mitigation in the future, due to any combination of biophysical and economic limits examined here, leaves us with no 'plan B'. As this study shows, *there is no NET (or combination of NETs) currently available that could be implemented to meet the <2 °C target without significant impact on either land, energy, water, nutrient, albedo or cost, and so 'plan A' must be to immediately and aggressively reduce GHG emissions.* (pp. 48–9 – emphasis added)

The worry voiced here is that the future use of NETs could serve as justification for the continued use of fossil fuels today, yet their failure leaves us with no Plan B. Nevertheless, the authors themselves seem fully aware of the political and economic obstacles to their warning being heeded when they reference the fact that with NETs comes a promise: things need not change now since we have big, low-carbon plans for the future – a Faustian bargain par excellence that weakens present-day climate ambition with assurances that a green future is just around the corner. As Anderson and Peters note in the journal *Science*, 'negative emissions technologies are not an insurance policy, but rather an unjust and high-stakes gamble' (p. 183). In this, the risk that comes with assuming, as the Integrated Assessment Models do, that the large-scale deployment of NETs mid-century will allow us to meet our mitigation targets is a risk rooted in 'moral hazard and inequity' (p. 183), since it is the world's most vulnerable and marginalized, and those least responsible for climate change, who will bear its worst effects soonest. Choosing NETs over aggressive and immediate mitigation in the near term maintains a status quo made possible by historic inequities between North and South, rich and poor, oppressors and oppressed, requiring thus that we keep our green-coloured glasses close at hand.

Solar radiation management
Distinct from NETs, solar radiation management (SRM) includes technologies that would block the amount of sunlight reaching Earth's surface in order to cool the planet. SRM is inspired by the experience of volcanic eruptions and their side-effects. In particular, a volcanic eruption sends into the atmosphere plumes of ash containing sulphur dioxide, which, once in the stratosphere, mixes with water to form sulphur aerosols whose properties include the ability to block solar radiation. The global cooling of around 0.6 °C that occurred in the year and a half following the eruption of Mount Pinatubo in the Philippines in 2001 was a natural example of SRM. Amongst other possibilities, SRM, used for the purpose of climate engineering, could include injecting sulphur into the stratosphere. While the technology remains largely theoretical at this point, there are those within the scientific and business communities that are actively pursuing the strategy to address catastrophic climate change. In the summer of 2017, a group of Harvard researchers with significant funding from Bill Gates and other individuals and foundations launched a study that plans to send small amounts of water, and then calcium carbonate, into the stratosphere, to analyse the outcomes as part of Harvard's Solar Geoengineering Research

Program. As with NETs, which, when analysed in isolation and free from the social, political, and economic landscapes which they are embedded within and interact with, SRM technologies possess the potential capacity to lower global temperatures to levels more consistent with pre-industrial times. However, as much of the scientific research on SRM has shown, there are significant risks, possible side-effects, and unknowns that would come with plans to manipulate the amount of solar energy that reaches Earth, prompting many to reject the technologies in line with the precautionary principle.

Some of the highest risks have to do with possible changes to global precipitation patterns that would come with SRM, potentially more significant in tropical latitudes.[10] Changing rainfall patterns have the potential to impact agricultural systems, food supplies, and thus food security for populations affected by the changes; with data pointing to the potential for more extreme changes to rainfall patterns in countries located in tropical latitudes, we can expect that some of the world's poorest and most vulnerable – those in Sub-Saharan Africa, Latin America, and Asia – may be disproportionately impacted by such changes. The other major risk that comes with SRM is that, once begun, the science suggests that SRM cannot be stopped without leading to a rapid warming of the global climate, possibly of up to 5 °C. This would require SRM to continue indefinitely, or otherwise threaten climate catastrophe. The potential for SRM to interact negatively with ozone – a major air pollutant that also damages crops – and the ozone layer is highlighted in the scientific literature. Questioning whether or not climate engineering, including SRM, is 'worth the risks and costs of its side effects' (Keller et al., p. 9), one group of scientists publishing in *Nature Communications* noted that their research suggests that the focus must be on the mitigation of CO_2 emissions rather than climate engineering.

Indeed, the risks of SRM are too great; while proponents, whether scientists or the business interests that fund them, argue that SRM is an option of last resort in the face of global inaction, or insufficient action, on climate change, and that SRM should complement, rather than replace, policies to reduce emissions, many of their colleagues in the scientific community are sounding the alarm bells about the possibility of deploying technologies that might bring with them devastating and catastrophic effects that would not be identified before it is too late. Research published in *Nature Climate Change* suggests that drought in the Sahel region of Africa is strongly influenced by sporadic volcanic eruptions in the northern hemisphere and the resulting impact on precipitation in that region. Citing data that suggests the death of up to 250,000 people, and the creation of some 10 million refugees, as a result of the Sahelian drought that spanned the 1970s–1990s, this research should prompt us to reflect on the moral hazard embedded in technologies such as SRM. If the data showed instead that large swathes of the United States and Northern Europe could be devastated by the side-effects of SRM, would we find then that the risks were deemed too great? Indeed, the debate on SRM exposes some of the most uncomfortable issues that lie at the heart of these technologies, particularly when viewed through the lens of climate justice and global inequality. Like NETs, much of the science on SRM shows that the uncertainty, risks, and limits to Plan B should demand of us an even greater resolve to pursue the Plan A of aggressive emissions reductions and the shift away from fossil fuels in the here and now.

Natural gas

And yet, as we move to consider our next possible pathway, that of natural gas, which is pitched as a 'bridge fuel' that will get us from our current high-carbon energy system to one that will be clean and green, we see that it involves governments and companies around the world investing heavily in fossil fuel exploration, extraction, and combustion.[11] The bridge fuel argument revolves around the fact that natural gas, when burned, produces between 50 and 60 per cent less CO_2 emissions than burning coal, meaning that, at the point of combustion, it has a lower carbon footprint. Advocates for natural gas in general, and for the idea that it should play a central role in achieving our emissions reduction targets, couple this data with the argument that natural gas is a reliable energy source to buffer against the intermittency of renewables, suggesting that any reduction in emissions is a good reduction in emissions – this in spite of the fact that we have the technology today to pursue no-carbon energy, and technologies to address renewables intermittency are rapidly developing in line with their growing economies of scale and cost-competitiveness relative to fossil fuels. Not surprisingly, some of the world's largest and most powerful fossil fuel companies back the bridge fuel argument, since it protects and advances their business model, and thus profitability, in a world where the science tells us that it is time to turn away from fossil fuels. In November 2016, a group of ten of the world's largest oil and gas majors unveiled the Oil and Gas Climate Initiative, their plan to contribute to the Paris Agreement (note that it was announced at a press conference on the day the Paris Agreement went into effect), which focuses primarily on advancing natural gas over coal, with additional support for CCS. With finances of $1 billion, and no renewable energy strategy, critics have highlighted the business-as-usual nature of the initiative, whose primary concern is maintaining market share for its backers.

The backdrop to this story is one in which natural gas has been pivotal in helping to revolutionize the global energy system, alongside the geopolitics that underpin it. The numbers globally tell part of the story – in 1990, natural gas was responsible for 15 per cent of global electricity generation, rising to 22 per cent in 2014, but it is in the United States where the significance of natural gas is truly revealed. Between 2006 and 2016, the share of energy provided by natural gas in that country increased by 54 per cent, beating coal in 2016 as the main energy source powering the nation. Much of this increase can be traced to the shale revolution: the development of technologies including horizontal drilling and hydraulic fracturing – the latter of which blasts water, sand, and chemicals into shale, or tight rock, formations deep underground, within which natural gas and oil are trapped, fracturing the rock to liberate the hydrocarbons. Stores of natural gas and oil that had previously seemed inaccessible were now made available, with US production of gas and oil increasing dramatically as a result. As successive US administrations have worried about the nation's dependence on imported sources of energy, particularly in the wake of the 9/11 attacks, the technologies associated with the shale revolution have set the nation on course for possible energy independence in the not-too-distant future, while turning global energy markets upside-down as new oil and gas flooded the market, depressing prices. Trump has made it a top priority to continue aggressively to advance the power of the US oil and gas sector; in the case of natural gas, the IEA reported in July 2017 that the USA is now poised to become one of the world's top exporters of the fuel, facilitated by tens of billions of dollars of investment in liquefied natural gas (LNG) infrastructure that converts the gas to a liquid, improving efficiency in shipping. In June and July of 2017, Poland and the UK received their first shipments of US LNG. Indeed, there are very powerful political economic interests backing the rising star of natural gas, with Shell CEO Bill Van Beurden asserting that, not simply a transition fuel, natural gas is a 'destination fuel' (quoted in King).

Yet the science has made it clear that this is a destination we must avoid. Most importantly, natural gas is a carbon-intensive fossil fuel. Although its burning may emit less CO_2 than burning coal, it emits CO_2 nevertheless, while our Paris targets show that we must reach net zero emissions by mid-century. Net zero is another way of saying that any carbon emitted will have to be balanced out with an equal amount of carbon removed, bringing us back to NETs and related technologies. Given the preceding discussion on the limits and risks of these technologies, investment in zero emissions technologies is what is really needed. One of the many problems with investing heavily in natural gas exploration, extraction, shipping (including pipelines), processing, and power generation is that it contributes to the extensive development of natural gas infrastructure that then risks the 'lock-in' of this carbon-intensive fuel and the emissions that result from it. 'Lock-in' here refers to a form of path dependency, so to speak, whereby the capital investments made in fossil fuel infrastructure require in turn that investors achieve a sufficient return on investment. This return can take many years to begin to materialize, with the infrastructure expected to remain in use, often for decades. This is the same for all fossil fuel investments. In short, whether a government or a private investor, within the current economic context, it makes no sense to sink massive amounts of capital into new infrastructure if it must be abandoned a few years down the road. The kinds of investment we are seeing in natural gas globally – and oil and coal, for that matter – send a sharp signal regarding the disconnect between our climate needs and those of the fossil economy, revealing the extent to which our Paris commitments are not being taken seriously.

Beyond the risk of lock-in, the science shows that natural gas, when full life-cycle GHG emissions are accounted for – so not simply those from burning it but also those that occur in extraction, shipping, processing, and so forth – is more potent than coal, the dirtiest and most carbon-intensive of the fossil fuels. Much of the reason for this stems from the fact that natural gas systems emit methane, a highly potent greenhouse gas. Recall our discussion of Global Warming Potentials (GWPs) from chapter 3, whereby different GHGs are assigned a value relative to CO_2 in order to assess their planet-warming potential compared to carbon dioxide. For methane, a short-lived climate pollutant that has an approximate atmospheric residency time of twelve years, the IPCC reports a GWP of 86 when calculated over a twenty-year timeframe. In other words, 1 tonne of methane emitted is equivalent to 86 tonnes of CO_2 emitted, meaning that the atmospheric impact of methane is far greater than that of CO_2 in the short run. Scientific research on the issue of natural gas, methane, and climate change demonstrates that the methane emissions resulting from natural gas systems, which are higher from shale gas than from conventional gas, are cause for serious concern, particularly when natural gas is framed as a bridge fuel. While it may be less *carbon*-intensive, its methane emissions make it potentially much more dangerous in the short term. Additional research has examined the consequences of expanding natural gas systems in place of near-zero-emissions energy systems – so, choosing to expand natural gas infrastructure instead of putting that investment into solar, wind, and other low- to no-emissions energy systems; the findings suggest that investment in natural gas will delay a more serious, non-fossil-fuel-based transformation of the energy system. These delays risk offsetting any potential climate benefits accruing from a coal-to-gas switch, given their role in impeding deeper structural change, especially given the problem of methane in natural gas systems.[12]

Natural gas illustrates why we cannot separate out carbon as though it can be dealt with in isolation. Specifically, shale gas is expected to account for 50 per cent of natural gas production in the USA by 2035, up from just 3 per cent in 2005. It is expected that natural gas obtained from tight sands, in combination with shale gas, will account for 70 per cent of natural gas production in the USA by that time. Globally, countries around the world have indicated their desire to expand the development of their shale gas industries, including China, with potentially the largest shale gas reserve worldwide. Both shale and tight sand extraction use hydraulic fracturing, more popularly known as 'fracking', to access and release the gas. Sufficient research now exists to show that chemicals used in fracking, many of them toxic, can contaminate fresh water underground, often sources of drinking water for nearby residents and communities. Researchers note that contamination of this nature is extremely difficult to investigate, given the complexity underlying it, pointing to potentially irreversible consequences, particularly when the cost of clean-up may be prohibitive. Methane leaking into water supplies has been documented, leading to social movement mobilization in many of the places where fracking is taking place, or has been proposed. Known as water protectors and water defenders, these movements oppose the expansion of an industry, along with the pipelines and infrastructure connected to it, that threatens their livelihoods in the name of an agenda that will further lock-in fossil fuel dependence well into the future. Studies also confirm the relationship between fracking and fracking-induced earthquakes, documented in Canada in the provinces of British Colombia and Alberta, and in US states where this form of drilling is common.[13] Indeed, when the above factors are taken together, the case against natural gas as a bridge fuel is exceptionally strong.

We can now return to the central questions to be asked of any potential pathway that is supposed to support a low- to zero-carbon transition. Recall that the first asks whether the pathway is effective in lowering carbon dioxide emissions in line with the science of climate change – although, as is clear now, we should add GHGs in general to that question. The second asks to what extent specific pathways fortify, or challenge – or both – the current world order and existing relations of power. Carbon capture and storage, negative emissions technologies, and solar radiation management are not effective strategies for lowering CO_2 emissions, since theirs are strategies that can allow for continued emissions, seeking to capture and store them after the fact, or to block out the sun's solar radiation so that simultaneous global cooling might take place. As such, they do not challenge existing relations of power or the global inequalities produced by the current system of fossil capitalism; yet each poses a serious risk related to land, water, and food security, amongst others, sure to impact disproportionately the world's most vulnerable populations. The same can be said of natural gas, whose expansion as an industry fortifies the power of oil and gas interests, charting further a pathway laced with fossil fuels. While some may suggest that it is possible to regulate some aspects of the methane problem that comes with natural gas – as President Obama sought to do with regulations that required oil and gas companies to capture methane at drilling sites, while upgrading equipment to prevent or reduce methane leakage and emissions – we need only to look at the Trump Administration's efforts to kill these regulations to understand the extent to which effective regulation relies on which way the political winds are blowing. Better to leave the gas in the ground and avoid the capital investments than to hope that the 'right' leaders will get elected. Indeed, none of the pathways reviewed so far promise socially transformative change, and we are thus left to look elsewhere, not to pathways paved with fossil fuels, but to those that open the possibility for real, low-carbon, and socially and ecologically just futures. It is to these openings that we now turn.

Where the grass may be greener

As the sub-heading suggests, whether a pathway will lead us to low- or no-carbon futures, rooted in social relations that are socially and ecologically just, is contingent. Contingency here has much to do with the kinds of technologies that are employed to get us there, but technologies are always deeply embedded within particular social, political, and economic contexts, as shown above. While it is important for us to know what technologies exist, what they might offer, and under what conditions their uptake and diffusion become possible, we must also ask important questions regarding who owns and controls them, and who might access them and in what ways. As Scoones, Leach, and Newell make clear in their analysis of the politics of green transformations, there is nothing inherent to transitions – they may maintain existing relations of power and privilege, they may create powerful new actors and inequalities, or they may open spaces to improve equity and justice. There is a strong moral and ethical argument in favour of supporting green transitions that are both equitable and just, however, particularly given the extent to which carbon is profoundly implicated in global injustice, including the injustice of a changing climate and its historical origins. In transitions, we have an opportunity to correct some of the wrongs, but this requires serious thinking about how we might do that. It requires paying attention to the complexity, contingency, and 'socio-technical dynamics of low-carbon transitions', and understanding how they are shaped by existing social and power relations, systems of production, vested and competing interests, unequal resources, entrenched cultural practice and beliefs, and institutional configurations. And it requires considering concrete examples for the lessons they offer.[14] Before looking more specifically at these questions, it is worth getting up to speed on renewables, since they do indeed dominate the policy landscape and much of the popular understanding of how we are going to tackle the climate crisis.

Renewables
Unlike fossil fuels that took many millions of years to form deep underground, renewable energy is energy that is replaced or replenished naturally, available for use by humans within timescales that match our own existence. Solar, wind, geothermal, tidal, hydro, and biomass represent the renewable energy sources that we tend to hear the most about, and for which global data is increasingly available.[15] According to data published by the IEA in 2016, renewable energy made up 13.8 per cent of the world's total primary energy supply in 2014. Of that, biofuels and waste accounted for 10.1 per cent, hydropower 2.4 per cent, and wind, solar, geothermal, and tidal 1.3 per cent. The majority of biofuels come in the form of solid biofuels /

charcoal, used non-commercially throughout the global South in particular, for household cooking and heating. The latter are not zero-carbon renewable energy sources, but the emissions associated with household cooking and heating in the global South from the burning of trees and biomass, for example, are known as subsistence or survival emissions since they emerge from activities required to meet livelihood needs. These can be contrasted with luxury emissions associated with mass consumerism, far more problematic when considered through the lens of ecological justice.

Recent data also shows that renewables, excluding large hydro, accounted for approximately 11.3% of generated electricity globally in 2016, compared to 6.9% in 2010, and 10.3% in 2015. Of 2016's new electricity generating capacity added across the globe, renewables reached a new high with a 55.3% share, excluding large hydro. This growth reflects fairly dramatic declines in the cost of renewables over the last few years; during 2016 alone, the cost of offshore wind generation fell 28%, onshore wind 18%, and solar 17%. This helps to explain why, with the level of investment in renewables down 23% in 2016 from the previous year, total installed renewable capacity was up to 138.5 gigawatts, an increase of 11 gigawatts over 2015's 127.5. In terms of electricity generated from wind and solar energy, Europe continues to lead globally – in 2016, generating 306 terawatt hours (TWh) of the former, and 110 TWh of the latter. During the same year, Asia saw 229 TWh of wind energy generated, and 94 TWh of solar, while in North America the figures stood at 226 TWh for wind and 38 TWh for solar. By comparison, in 2016 South America generated 27 TWh of wind and 2 TWh of solar, Africa 8 TWh of wind and 4 TWh of solar, Central America and the Caribbean 4 TWh of wind and 1 TWh of solar, and the

Middle East 0 TWh of wind and 2 TWh of solar. Not surprisingly, the emerging economy powerhouses of China, India, and Brazil have led the way in renewables investment in the global South; in 2015, those three countries combined saw investments of $132 billion, compared to $145 billion across all developed countries, and $35 billion across all other developing countries.

While the growth of renewables is thus encouraging, though uneven, it remains important to keep in mind the magnitude of the task that lies ahead. With renewables supplying 13.8% of total primary energy supply in 2016, with a mere 1.3% of that coming from wind, solar, and geothermal, we are reminded of the fact that over 80% of the world's total primary energy comes from fossil fuels, with 4.8% derived from nuclear energy. Moreover, global energy demand will continue to grow, with projections suggesting an increase of almost 50% between 2012 and 2040, much of it originating in countries of the global South. Alongside these projections, we know that fossil-based energy infrastructure continues to proliferate globally. Recent data shows that G20 governments, despite their commitments under the Paris Agreement, provided $122.9 billion in public finance to fossil fuel projects between 2013 and 2015, not including investments and subsidies from state-owned banks and enterprises, which is almost four times the level of public finance going to clean-energy projects. As discussed in chapter 1, the IMF has calculated the value of fossil fuel subsidies globally per year at $5.3 trillion on a post-tax basis. This context creates a tremendously complex and multifaceted challenge since the ideal pathway would include the transitioning of existing fossil fuel infrastructure to non-fossil sources in conjunction with improvements in energy efficiency, and the development of new energy infrastructure that adds to overall capacity from renewable sources. Various researchers have modelled scenarios showing that a full-scale rapid transition to renewables is technically feasible,[16] so the question becomes one of the corresponding social conditions facilitating, or hampering, the process. Since ample space has been dedicated to discussing these social conditions throughout this book, the focus will now shift to renewables and low-carbon energy in practice, looking at the distinct conditions under which they might be, or are being, developed, so that we may assess their prospects for generating deeper transformations.

While the market price for low- or no-carbon energy sources such as wind and solar matter in terms of providing market signals to investors, the policy landscape is much more important in terms of providing a framework that requires a green transition. Low-carbon policies include regulations requiring emissions caps and/or reduction targets, low-carbon fuel standards, renewable portfolio standards, energy efficiency requirements, green building standards, and carbon pricing, to name a few. With the right policies in place, emitters, including producers and consumers, are forced to change behaviours in line with climate goals. On the other side of the coin, the transition to a low- to no-carbon energy system is going to cost a lot of money. In 2014, the IEA estimated that, between 2012 and 2035, investments of $53 trillion in global energy supply and energy efficiency would be required to limit warming to 2 °C; 1.5 °C would require substantially more. Carbon finance is thus a critical piece of the puzzle, yet not all finance is created equally – where it comes from and what interests it serves have profound implications in terms of pathways.

Various researchers make a distinction between 'impatient' and 'patient' finance or capital, noting that the former seeks to maximize profits as quickly as possible, leading to 'short-termism' in investment strategies. Impatient finance is often found in commercial and investment banks, portfolio equity and bond investors, and venture capital and hedge funds, with investors looking to get the most bang for their buck, as soon as possible. The highly irresponsible forms of lending that precipitated the subprime mortgage crisis in the USA in 2007 are extreme examples of impatient finance and its destructiveness. Patient finance, on the other hand, takes a longer-term approach to investment, expecting that returns may be slow in coming; it may also be guided by motives distinct from those of profit maximization, such as providing public goods and acting in the public interest, or supporting social, environmental, and ethical standards. Non-commercial financial institutions and instruments, including community and cooperative banks, national public development banks, and socially responsible and impact investors, are just some examples of patient finance. Importantly, impatient finance is often little interested in green or renewables investments that might be risky and that require longer timeframes to realize returns, which might be lower than those on other possible investments. Their private orientation also favours models of investment based on commodification. Water has been a perfect example in this respect – when the World Bank required countries like Bolivia and South Africa to privatize their water systems in return for badly needed loans in the early 2000s, the private companies that received the contracts, in seeking to maximize profits, began charging fees that were well beyond the reach of the poor, turning water into a private for-profit commodity, versus a public good without which humans die. Under such a model, there is little incentive to respond to the needs of the poor since there is little wealth to be extracted from them.[17]

Energy, and access to it, are similar in many ways, given their centrality to human reproduction and quality of life, with the term 'energy poverty' used to describe circumstances in which individuals do not have access to energy sources, either because they are not readily available, or because they are unaffordable. When seeking new energy pathways that can reduce carbon emissions while challenging existing relations of power and inequalities, models that would commodify energy resources for maximum profit, as has been the case with fossil fuels throughout history, will only reinforce the existing order, requiring thus that we look to low-carbon pathways that prioritize universal accessibility, and social and democratic control and justice.

Germany's Energiewende, or energy transition, is one of the most oft-cited examples of a pathway that, though imperfect, embodies many of these priorities. Recent data shows that approximately 47 per cent of installed renewable capacity in that country is controlled locally by communities, farmers, cooperatives, and municipal utilities, amongst others. In turn, roughly 33 per cent of Germany's electricity is generated from renewable sources. The stunning level of local ownership has emerged through a combination of factors, both political and economic. No doubt, early social movement organizing against nuclear power, and in support of various environmental goals in the 1970s and 1980s, was important to the growth of renewables, but so too was Germany's decentralized political system, alongside the technology of renewables that allowed for decentralized control: Does the sun shine on your roof? Does the wind blow in your local fields? Instrumental to the spreading and up-scaling of local efforts was, in turn, the 1991 adoption of Germany's now-famous Feed-in Tariff (FIT) – which provided renewable energy producers with guaranteed prices over twenty years, essentially eliminating the investment risk while promising returns over time – and its Electricity Feed-in Act, which required grid operators to buy renewable energy first, rather than going to non-renewable sources, on the spot-market. Finally, the adoption in 2000 of the Renewable Energy Act, with a 2050 target of 80 per cent of electricity from renewables, massively expanded the process already underway. In all of this, patient, state-driven forms of finance were essential to supporting the spread of community energy projects in a way that private commercial financiers would not, given the risk and lack of scale. Germany's state development bank, KfW, actively supports local projects and enterprises with preferential lending and rates, while its regional state-owned banks, or Landesbanken, actively finance community-owned renewables.

The result has been a model that has allowed local communities and coops to help build a movement that has been able to challenge incumbent fossil fuel interests; that has put control and ownership of energy systems more squarely into the hands of communities and with public utilities; that has allowed for local economic regeneration and value creation; that has allowed for greater democracy in decision-making; and that now supports over 400,000 jobs in the renewable energy sector. Of course, we must not valorize the local simply because it is local, or communities simply because they are communities – what we want to do is analyse the benefits in comparison to energy systems dominated by large multinational corporations and utilities with deep fossil roots. Indeed, it is this challenge that has led fossil interests to launch a powerful assault on the project with a corresponding public relations campaign that has worked hard to undermine it.

This should not surprise us – in challenging incumbent and powerful interests, with Germany's nuclear industry owing its death in part to Energiewende and the forces that support it, alternative pathways will face fierce opposition and are especially vulnerable. Energiewende's future is thus uncertain, with the government recently eliminating the twenty-year price guarantees while capping the growth of renewables. A sober assessment of Energiewende must acknowledge that, by embedding fixed renewable price subsidies within consumer electricity bills, the poor and lower-income citizens have been disproportionately burdened with the cost of the project. This fact has been used by private utilities and fossil interests in their campaign to undermine Energiewende, to which the government has responded. A critical lesson thus has to do with how the costs and burdens of financing pathways are distributed.[18]

Another noteworthy example comes from the province of Nova Scotia, Canada. In 2007, the provincial government passed the Environmental Goals and Sustainable Prosperity Act, requiring 18.5% of electricity to come from renewable energy sources by 2013, 25% by 2015, and 40% by 2020. In 2005, only 8% of electricity was generated from renewables, rising to 22 and 28% in 2014 and 2016, respectively – this in a province heavily dependent on coal. An important policy support for this transition came in 2011 with the introduction of the Community Feed-in Tariff Program, or COMFIT, under the government of the New Democratic Party. The first of its kind in North America, COMFIT offered guaranteed prices over twenty years for majority community-owned producers of renewable energy, including wind, tidal, hydro, and biomass. Designed to encourage local buy-in, eligible participants included First Nations, coops, municipalities, universities, non-profits, and Community Economic Development Investment Funds. With a target of 100 MW of electricity, COMFIT exceeded expectations, awarding 182 MW between 2011 and 2016, with its projects powering over 70,000 homes and making up 20% of installed renewables in the province. With over CAD$45 million in direct investments, $135 million in project spending locally, and the creation of many hundreds of jobs in a province whose population stands at 940,000, COMFIT's community focus helped participants to decentralize and democratize control over the electricity system in a province dominated by a private regulated monopoly. Moreover, many projects included a 1% community dividend derived from the sale of electricity to the grid, used to fund local community projects. Importantly, in a country where First Nations communities continue to face severe discrimination and marginalization, and where fossil energy projects continue to receive government approval on their lands without their approval, COMFIT is noteworthy for the amount of indigenous-owned renewable projects it supported. In total, Mi'kmaq indigenous communities own 40 MW of renewable energy projects under COMFIT, an amount that exceeds the amount of electricity used in the 13 Mi'kmaq-band communities in Nova Scotia.

The experience of COMFIT provides another powerful lesson on how a supportive policy context can open space for more transformative low-carbon transitions. Not only did it provide room to challenge deeply entrenched fossil interests in the province, but it prioritized an ethos of local collective ownership that challenged the power of large-scale commercial interests that tend to dominate in many renewables markets globally. Unfortunately, the Program was cancelled abruptly by the Liberal government in 2016, with the government blaming COMFIT for increased electricity rates in the province to justify its termination. In fact, between 2004 and 2013, the price of coal increased by 75%, significantly impacting electricity rates in a province where energy poverty is especially pervasive. Rate politics are thus intense in Nova Scotia, and while COMFIT did mobilize community support, its base is not nearly as far-reaching as that in Germany, underscoring the need for significant grassroots mobilization in support of favourable policies, which can respond forcefully once they come under attack.[19] Indeed, what this case helps to reveal about the politics of carbon transitions is that political institutions and histories matter. Nova Scotia, similar to all Canadian provinces, has a first-past-the-post voting system that frequently produces false majorities for governing parties. In decision-making, majority governments need not negotiate and compromise with other parties – features much more typical in systems with proportional representation. If citizen dissatisfaction leads voters to punish incumbent parties by voting them out, their policies are always vulnerable if they lack the kind of widespread support that would make a new government think twice before gutting them.

Many countries in the global South, while highly differentiated, face challenges distinct from those in the global North. Installed electricity capacity in Sub-Saharan Africa, home to approximately 1.03 billion people, for example, is approximately 77 GW; with the exclusion of South Africa, it is 33 GW. This compares to Germany's installed electricity capacity of approximately 200 GW. It is estimated that approximately 1.2 billion people globally lack access to electricity, 95 per cent of whom reside in Sub-Saharan Africa and developing Asia, where existing electricity infrastructure is often degraded and unreliable, located far from rural inhabitants. Over 23 million people lack access to electricity in Latin America and the Caribbean. The need to more fully electrify these nations, home to some of the world's poorest and most vulnerable populations, is significant. Leap-frogging fossil-fuelled electrification in many areas of the global South is a real possibility, given the abundance of renewable energy sources that need not, necessarily, connect to national or regional grids – but also given the drastic decline in the cost of renewables. Much of the development literature highlights the potential for mini-grids or off-grid renewables to meet the above challenges, noting the potential benefits these can bring to local communities in terms of democratic decision-making, local control over electricity, local jobs, and so forth. But communities, South and North, are diverse and differentiated social groupings cut through with power. The potential for some community members to capture the benefits of local renewables, and for marginalized groups to find themselves excluded – or, worse, their lands lost to service a particular project – is very real. For local renewables projects to be transformative, they require forms of organization and participation that facilitate equitable representation – a tall order, regardless of the location.

Equity and shared benefits are more likely when renewables remain in public as opposed to private hands, however. Evidence from Sub-Saharan Africa shows that, where private entities and market-based approaches to mini- and off-grid renewable provision have been encouraged, private companies tend to charge higher rates in the interest of commercial viability and profitability, a serious problem when servicing poor communities, as noted above. Terms of finance also determine who is able to participate in the development of renewables capacity. When the World Bank's International Finance Corporation launched its Photovoltaic Market Transformation Initiative in 1998 in Kenya, a project intended to expand solar home systems throughout the nation, the initiative shut out many Kenyan companies, since a minimum investment of $0.5 million was required. This was well beyond the reach of many

companies, communities, coops, and non-profits, who had
no hope of accessing the initiative. Such groups wanting to
establish mini- and off-grid renewable generation are confronted with high and often prohibitive costs which private
banks are uninterested in financing. The public sector has
a significant role to play in developing these systems, and/
or in financing capital costs, offering FITs through which
communities can feed excess capacity to national grids, and
ensuring that long-term maintenance, replacement, and
upgrading costs can be covered. Grant-financing, preferential finance from state development banks, and multilateral
financing favourable to local ownership is thus critical,
as is a policy context that welcomes and rewards affordable micro-generation. Moreover, in cases where national
public utilities can advance equitable generation and distribution, they should be encouraged – in this, there is
nothing inherently superior to local rather than national
approaches, *per se*. Nevertheless, with multilateral institutions like the World Bank, regional development banks,
and Northern development agencies prioritizing market-
based, private approaches to electricity generation in the
global South, alongside state-based neoliberalism and previous rounds of power market liberalization, and in the face
of incumbent fossil interests in many nations – including
the state, domestic and foreign owners of capital, labour
movements in the fossil fuel sector, and households that
may benefit from relatively cheap fossil-generated power
– the challenges of achieving equitable and publicly driven
electrification remain extensive. This is the case for low-
carbon energy transitions in general, particularly when
many Southern nations – South Africa, Nigeria, Venezuela,
Trinidad and Tobago, and many more – are themselves
heavily involved in fossil fuel extraction, processing, and
distribution to meet domestic energy needs, and/or to

export to foreign markets. These contextual realities and political economies underscore the need for sufficiently strong democratic mobilization in favour of alternative projects which can align with coalitions of interests seeking to advance this agenda.[20]

Many observers point to Costa Rica and Uruguay to underscore the importance of state-led and publicly owned renewables transitions. In Costa Rica, 99 per cent of electricity and 79 per cent of energy is derived from renewables, with nearly universal access to electricity. With no private-sector involvement, the electricity sector is controlled by four state-owned companies and four community-owned non-profit energy cooperatives, the latter of which boast a total membership of 180,393 members, servicing 392,071 users, and providing over 1,960 jobs. Costa Rica's history of social democracy and welfare-state politics dating from the 1940s helps to explain its energy trajectory, with the state actively pursuing socially progressive policies that prioritized public, democratic control. This also led to the majority of the banking sector remaining public in Costa Rica, with the nation's Banco Popular an example of a bank that is worker-owned and controlled. This opens the door to the kinds of patient financing models that are much more conducive to transformative low-carbon transitions. In Uruguay, where universal energy access has been secured, 94 per cent of the country's electricity and 55 per cent of its energy comes from renewables. This has been attributed to active government planning, with the national power company, UTE, remaining in public hands. Observers note that, in ten years, Uruguay went from having no installed wind capacity to being the nation with the highest proportion of wind in its energy mix worldwide. Both models face challenges. Costa Rica's heavy dependence on hydropower leaves it vulnerable to the changing rainfall patterns and drought that are accompanying climate change, pushing local actors to expand the renewables portfolio in response. In Uruguay, the state-owned petroleum company is undertaking exploratory drilling in partnership with Exxon and others. Moreover, critics suggest that, while admirable, levels of popular participation in both cases remain insufficient, signalling the need for greater levels of grassroots mobilization to defend and expand the benefits of these projects.[21]

Indeed, while an active push for renewables is occurring throughout much of the world, with countries such as China implementing aggressive policies and renewables requirements domestically, committing significant finance to the development of a renewables industry that is now able to out-compete most others globally, and that sees its generating companies expanding into markets around the world, it is clear that the renewables 'revolution' is uneven, and its benefits are highly contingent. Existing political economies, energy geographies, and political and institutional histories, and how citizens fit into all of this, shape who controls and who benefits from the roll-out of renewables, and thus their emancipatory potential.[22] What is clear is that there is no silver bullet for this process, although the discussion has sought to identify which pathways are more desirable when the end goals are those that combine low- to no-carbon energy systems with socially and ecologically just transformations. What we know is that, while transitions are under way, they are only just beginning, meaning there's still much work to be done in thinking through what these examples might teach us, and what obstacles they might face.

The next chapter

How to turn the page so that our next chapter on this planet might be one that can sustain, with justice, present and future generations of human and non-human life is where we are headed next. The challenge is momentous and, at its core, it revolves around issues of democracy and justice in a world that remains dominated by fossil power. It is this power that continues to shape profoundly how we live and experience our day-to-day lives, and how we understand the world around us and where we fit into it. If we are to rise to the challenge, then surely the time has come to remove our green-tinted glasses. Indeed, if there's another lesson to be drawn from *The Wizard of Oz*, it is that, once the Wizard himself was unmasked, and the illusion of the Emerald City and his perceived power debunked, Dorothy and her companions were able to realize that the power to achieve their desires and to change was within them all along. Change, if it is going to come, can and must come from us. We cannot wait for our Wizard, promising future riches bathed in green, to act on our behalf. It is to this challenge that we turn in the final chapter of this book.

CHAPTER FIVE

The Future of Carbon Politics

In 2017, an issue of the *Journal of Political Ecology* included a Special Section on 'Political Ecologies of the Green Economy'. The contribution from anthropologist Sian Sullivan, 'What's Ontology Got to Do with It? On Nature and Knowledge in a Political Ecology of the "Green Economy"', proved especially useful in thinking through a number of the themes covered in this book. Not only did it interrogate this thing called the *green economy*, now ubiquitous in policy discussions on how to solve the world's overlapping environmental crises, including climate change, but it also dug deeper into the problematic of truth, knowing, and power. Green economy discourse offers a narrative within which, once nature and the services it provides are valued as market goods – natural capital – set on equal footing with all those other things whose economic value is well established – oil, gas, lumber, fish – nature will have a fighting chance. In other words, once our eyes, or pocket books, are opened to the economic value embedded within the critical ecosystem functions performed by nature – so forests functioning as lungs, inhaling carbon dioxide and exhaling oxygen – we'll then have a powerful incentive to save it. Sullivan notes that ontology, focused on the study of 'what is' or 'reality', reminds us that *reality* is not settled, that our ideas about the world and how we know it, and how we should act in it, are contingent, carrying with them different ethical commitments. Yet narratives of the green economy would have us believe otherwise: that there is one truth, one way of knowing, and one way of acting, determined largely by the logic of capital and its social relations. Sullivan begins her article with reference to Marion Zimmer Bradley's novel *The Mists of Avalon* (1982), within which the legend of King Arthur is retold from the perspective of the women who are central, though always secondary, to Arthurian lore. It is a tale of two worlds: one where the priestesses of Avalon and the Lady of the Lake uphold the power of the Goddess and pagan ritual, shrouded in mist and increasingly unreachable and unseen by those living in the other, dominated by an ascendant Christianity and its patriarchal mores, seeking to banish all but God's one 'truth'.[1]

Having read *The Mists of Avalon* as a student in high school, I was inspired to read it once again, particularly given the connections drawn by Sullivan between the struggle, ultimately futile, waged by the priestesses of Avalon to remain of the world when confronted with the power of a singular Christianity, and contemporary struggles to resist filtering all that we do and all that we see through a narrow economic rationality. In the following passage, Morgaine, the novel's main protagonist, seeks passage through the mists to Avalon, but is told that, in the shadow of the Christian church, passage has become increasingly difficult, prompting her to reflect on belief and its capacity to construct the world and all around us:

> She said to one of the men, 'Can you take us to Avalon? Quickly?' He said, shivering, 'I cannot lady. It grows harder, without a priestess to speak the spell, and even so, at dawn and at noon and at sunset, when they ring the bells for prayer, there is *no* way to cross the mists. Not now. The spell no longer opens the way at these times, although, if we wait till the bells are silent, it may be that we can

manage to return.' Why, Morgaine wondered, should this be so? It had to do with the knowledge that the world was as it was because of what men believed it was . . . year by year, these past three or four generations, the minds of men had been hardened to believing that there was *one* God, *one* world, *one* way of describing reality, and that all things which intruded on the realm of that great one-ness must be evil and of the fiends and that the sound of the bells and the shadow of their holy places would keep the evil afar. As more and more people believed this, it *was* so, and Avalon no more than a dream adrift in an almost inaccessible other world. (p. 749)

I follow in Sullivan's footsteps here, given the power of the analogy. Green economy discourse is but one facet of a larger process underway in the contemporary world. It is one in which neoclassical economics and its neoliberal applications are the new religion, giving us one truth, one way to know the world, one way to be in the world. In this, 'nature' is reconceptualized in such a way that its ultimate worth is to be determined in a market that sees no difference between coal, tar sands oil, bombs, and coral reefs, and that demands payment from all, regardless of means, to access its goods. Within this singular world, moreover, our own subjectivities as sentient beings on this planet are progressively narrowed, our engagements with non-human forms of life and with each other mediated ever more so by and through the commodity form and a particular kind of economic rationality. We see these battles emerging with respect to carbon. Many of the efforts to deal with too much carbon isolate it and reduce it to its molecular form – CO_2. From there, it becomes easy to count and easy to commodify, stripped of its complexity and sociality.

What is particularly useful in the above quote is the presence both of structure (Christianity/religion) and of agency (the individual). It makes clear that the power of the structure is vast and encompassing, pointing to the role of individuals and their belief systems in fortifying that power. At the same time, structure is not absolute; nor are human beliefs – as the tide of Christianity continued its forward surge, there remained those who challenged it, refusing to let other worlds forever slip into the mists.

This book has thus far dedicated a significant amount of space to exploring the structure of what remains a largely fossil-fuelled capitalist economic system, and its role in the climate crisis. It is a structure that embeds most of us, to varying degrees, in complex carbon networks and dependencies shaped according to historically rooted patterns of growth and development, making it particularly hard to call for other worlds not based on economic growth and the attainment of profit. It is economic growth that, for the lucky few, provides jobs and income to nourish, clothe, and shelter their bodies. For the lucky few, it is economic growth that secures their pensions, their health insurance, their ability to pay school fees, and more. For the unlucky, living in a world that becomes more commodified by the day, the hope of one day entering this system is fierce. Indeed, our material imperatives as human beings – to eat and sustain our bodies – demand our participation in this structure. And it is a structure so normalized, at this point in time, that its obscenities and oppressions appear external to its logic. Despite the fact that we live in a world where food is treated as a commodity, where more than enough food is produced every year to feed the entire global population, where an estimated one-third of that food – approximately 1.3 billion tonnes – is wasted annually, where an estimated 795 million people between 2014 and 2016 suffered from chronic undernourishment globally according to the UN's Food and Agriculture Organization (FAO), and where an estimated 9 million die per year from hunger, far too few question the structural violence that is hunger and starvation, so normal has it become. Despite the fact that we live in a world of obscene inequality – according to Oxfam, in 2016, just 8 men possessed the same wealth as the world's poorest 3.6 billion people combined – too few question the structural violence wrought by an economic system that normalizes this level of extreme disparity. And despite the fact that we have ample evidence showing the clear relationship between our current economic system and the global climate crisis, wherein the scientists have shown beyond a shadow of a doubt that it was with the onset of the industrial revolution and global capitalism that levels of CO_2 in the atmosphere began their alarming increase, with concentrations now sitting beyond 400 ppm, too few really and truly question the nature of capitalism as an engine driving the collapse of ecosystems and environmental services the world over. The climate crisis is but one in a series of overlapping global ecological crises – the collapse of fisheries globally; our entry into what is being called the world's sixth mass extinction, with massive and widespread biodiversity and species loss; and the list goes on.

Despite all of this, when voices emerge questioning the nature of capitalism as a system of production whose growth imperative and profit motive all too often run counter to the interests of humanity, global ecology, and social justice, those voices are pushed further to the margins – the present-day mists. Accused of blasphemy and heresy against the God of economic growth and economic rationality, those voices are discredited for lacking in realism, for being too radical, for blocking real progress. If successfully silenced, all that is left is the ringing of the church's bells.

Unlike previous chapters, this chapter re-scales the discussion to focus more closely, though not exclusively, on the individual, with an eye to questions of structure and agency in an era of catastrophic climate change. It attempts to respond to that so often uttered question, 'But what can I do?', for that question, and how it is acted upon, are shaped profoundly by structural factors. Moreover, it impinges deeply on who we are, who we perceive ourselves to be, and what kind of futures we imagine for ourselves. In this, we will once again return to the question of pathways in order to explore how and why individuals set out on some pathways, but not others. We will ask who determines the route, whose interests are served, and the requisite identities and subjectivities of its travellers – are we consumers or citizens in our travels? What will ultimately be argued is that only by travelling as citizens who mobilize politically and collectively for democratic change and climate justice, requiring thus that we challenge the hegemonic truths of advanced fossil capitalism and its grip on states around the world, will we have a chance of limiting the climate crisis for the benefit of the planet and the majority of its inhabitants.

Who am I? Catastrophic climate change and the role of the individual

Concerning consumption
In 2010, the artist Gotye released the song 'Eyes Wide Open', whose narrative suggests the end of days for humanity, a humanity defined largely by its drive to consume, destroying all that we've achieved. It is a powerful song rooted in what seems to be a deep frustration – we see what is happening but we won't change. The reason for this, the song suggests, is that the very essence of humanity is that of the consumer – we're little willing to part ways with who we truly are. When I first heard the song, and then listened to it many times over, I appreciated its contribution, especially

in a popular-culture forum, to this critical issue. At the same time, however, I felt unease, particularly with how an undifferentiated humanity is framed and defined, and with the suggestion that the blame sits squarely at the feet of the individual – the consumer. In fairness, Gotye sings of a problem that is stunningly destructive – the overconsumption of the planet's resources, including carbon-spewing fossil fuels, so that Earth's life-support systems are now crumbling under its weight. Overconsumption is distinct from consumption, however, with the World Bank noting in 2008 that the world's richest 20% were responsible for 76% of private consumption globally; the middle 60% were responsible for 21%; and the world's poorest 20% were responsible for a mere 1.5%.[2] Overconsumption is rooted deeply in global inequalities, with the ability to consume highly dependent on one's class, race, and gender, amongst other things.

Moreover, as this book has emphasized throughout, a fundamental system requirement of our current mode of production is that of consumption – without it, growth and profitability are not possible. For those who study the political economy of overconsumption and capitalism, a typical starting point for analysis is the rise of the Fordist system of production in the United States, pioneered by Henry Ford through the Ford Motor Company in the early 1900s, and that, with the introduction of the assembly line, permitted mass production, for which mass consumption was required – why else produce all those cars? With an initial focus on consumer durables, like Ford's Model T car, home appliances, furniture, and so forth, it soon became clear that more products were required to sustain the growth of the economy. Alongside durables, non-durables began to proliferate, all requiring continuous consumption lest the system collapse. Having undergone various transformations in production strategies, it remains the model of mass production tied to mass consumption that has proliferated around the world, aided by neoliberal globalization and the ability of capital to produce where it is cheapest as a strategy to increase profitability further.

What this history shows us is that overconsumption emerges within a particular structural and political economic context; the act of over-consuming is not *a priori* to this context, somehow inherent to the human condition. Yet, within this context, every effort is made by those who need us to buy their goods, and who profit from this, to define us as consumers and to have us self-identify as consumers, so that our existence as consumers is normalized, beyond question. Does this mean that people don't enjoy consumption? Of course not. In many ways, the sea of goods on offer under capitalism taps into human desires for pleasure, for symbols that define and give us meaning, and for status. They also tap into something else, and that is an enduring belief, for many, in what we are told capitalism is supposed to be. Capitalism itself emerged at a time when ideas of progress, science, and modernity were ascendant – it is inextricably tied to their histories. Together, they promised freedom from oppression – a nature beyond our control and to whose domination humans were forced to submit throughout most of history; feudal political systems that repressed human ingenuity, liberty, and individual rights; and much more. There is something deeply threatening, emotionally and existentially, in the idea that perhaps we, as a species, haven't triumphed, that we remain at the mercy of nature, that democracy and freedom might be shams, and that science might not save us.[3] This combination of structural hegemony (cue the church bells), pleasure, and fear, to varying degrees, means that sticking with the status quo not only is easy, it is affirming, reassuring, and satisfying – our eyes may be wide open, but our field of vision is multi-dimensional. Seen through the lens of this particular conjuncture of forces, or factors, we are better placed to analyse the appeal of pathways that require us to travel as consumers.

The reusable shopping bag
Let's take the reusable shopping bag as a case in point, drawn necessarily from spaces where overconsumption is ubiquitous. Many readers of this book are likely to have a reusable shopping bag that they take to the market, grocery store, or shopping mall when running errands. In fact, it is likely that the reader will have many of them – five, ten, twenty, more? The idea of the reusable shopping bag emerged in response to the disposable, single-use, plastic bag, and its heavy environmental footprint. Indeed, the idea that we could now reuse our bags over and over instead of filling more landfills with plastics that break down and contaminate the environment became highly appealing. Unfortunately, it is not nearly so simple.

While businesses involved in plastics may not love reusable bags, they fit well within the broader logic of the capitalist system. Rather than targeting the problem of over-consuming, they allow us to continue, with only a minor tweak to the method – buy as much as you want, but, *please*, put it in a different bag. Nested within the world of consumption and its methods, this pathway allows us to travel as 'consumers'. While people do travel multiple pathways, we know that the discourse of green consumption is particularly powerful, with many slotting their 'eco' activities into that realm alone. Critics note that the discourse of green consumption identifies individuals and their personal choices as the cause of environmental crisis, not an economic system requiring limitless growth and consumption. Diverting our attention thus, the discourse prompts us to act as individuals rather than collectively, since collective action poses significant threats as well. All of this is aided by contemporary neoliberal politics that emphasizes the superiority of the market, and individuals participating in it, over the state, to achieve the greatest good. In this vein, the voluntarism of green consumption helps to keep government regulation at bay, suggesting that the tide of consumer and company support – say, for reusable bags – proves that we are capable of addressing the problem on our own. Absent from this framing is any acknowledgement of how the intersections of class, race, and gender shape who the green consumer is – the price tag for 'green' products remains well beyond the reach of many. At the level of the individual, there is something deeply comforting in the idea that solving the environmental crisis can be this easy, promising not hardship but continued pleasure. Unfortunately, we're not going to save the world through individual, voluntary actions. How many people do you see still shopping with plastic bags, after so many years of information on the good that can be done with that one simple choice? And here is where it gets even more complicated. While we could explore why it is that some do not seem to care, it is elsewhere that we are going.

Did you know that a reusable cotton bag would need to be used 131 times to match the Global Warming Potential (GWP) of a plastic bag used only once? Many reuse plastic bags as garbage liners, or for other purposes. Your cotton bag would need to be used 327 times to equal the GWP of a plastic bag used twice, and 393 times if the plastic bag is used three times. Recall that, when used for greenhouse gases, GWP measures the warming potential of each gas against 1 tonne of CO_2; the higher the number, the more potent the greenhouse gas. When applied to bags, we see that cotton bags are much worse for the climate due to the resources used and the nature of production – unless, that is, one makes sure to use them often enough to tip the balance. As reusable bags have proliferated, their durability has declined, resulting in a decrease in their GWPs – one would need to reuse a bag made from recycled polypropylene plastic 11 times to match the GWP of a plastic bag used once, and 26 and 33 times for a plastic bag used two and three times, respectively. These findings were presented in a report published by the UK Environment Agency in 2008 that examined the production, use, and disposal of different kinds of bags, along with additional measures of environmental impacts. Shipping was not included and it is unclear how the numbers would change if it was, though we do know that the supply chains of plastic and reusable bags are often extremely long, starting in low-wage manufacturing sites and then shipped to the country and point of sale.[4]

Adding another layer of complexity, companies soon saw the advertising potential in the reusable bag. Go into almost any retail store today and you will find racks of reusable shopping bags with company logos emblazoned on both sides. The branding genius of this strategy has led to countless companies now giving away reusable bags for free with a purchase. This means that many people have far more reusable bags than they need, complicating the picture further – as the number of bags owned by each individual increases, the less likely they are to use them sufficiently to offset their global warming impact. By way of anecdote, my regular grocery store advertises itself as the first plastic-bag-free store in Atlantic Canada. When customers forget their reusable bags – a not infrequent occurrence – they must buy the branded bags on sale for $0.99 at each cash register. For large purchases, that may mean four, five, or six bags – all destined to join those left at home.

Is the point of this discussion to advocate for plastic bags? No – the plastic bag companies have that one taken care of, and there is significant evidence demonstrating the environmental destructiveness of plastic bags. Measuring a cotton bag against a plastic bag does not mean that plastic is environmentally benign, just as measuring methane against carbon dioxide does not mean CO_2 is benign. The point of this discussion is, rather, to shed light on the many problems that come with pathways that channel action through a consumer slot that does nothing to challenge capitalism's fundamental imperatives of growth and consumption, and the high-carbon lifestyles that come with them. The reusable bag allows us to think we are doing something important – although when we scratch the surface, we see how problematic that assumption might be – while continuing to sanction consumerism, since it is through the act of consumption that we become 'green', the two fused conveniently. Since it also lacks a commitment to deep transformation, it is not difficult for capital to seize on the opportunities in green consumerism to grow further, to expand profit margins even more. While the reusable bag companies are doing well, our bodies become vehicles for further consumption and commodification – we are walking advertisements, further normalizing lives lived as consumers. And hey, if it's really that easy, we needn't face the music – we can continue to indulge as though there are no limits.

Perhaps most problematic, the implication here is that, if the solution must come from consumers, it is consumers that are the problem. There is no better way to protect a system of production than by shifting blame onto a self-interested and over-indulgent consumer – producers are powerless, responding simply to what consumers want. Of course, the narrative is not black-and-white; consumers do want things and producers produce in a system with very clear rules laid out for them, but in much more complicated ways, shaped deeply by existing relations of power. But once we are convinced that we are the problem, as consumers, or individuals – the two often used interchangeably – the less likely it is that we might look elsewhere for other sources, other answers.

Going it alone?
The proliferation of consumer and individual solutions to climate change is dizzying. A recent study that gained international media attention analysed the top lifestyle choices people can make to reduce their individual carbon emissions, looking specifically at developed countries, given their high carbon intensities. High-impact choices were then considered against government documents and school textbooks in select countries (textbooks were drawn from seven Canadian provinces), to see whether or not high-impact choices were typically recommended via these mediums. Of the list of twelve top lifestyle changes, seven fell into the high-impact category, four into the moderate-impact category, and one into the low-impact category. Topping the high-impact list was the recommendation to 'have one fewer child', followed in descending order of impact by 'live car free', 'avoid one transatlantic flight', 'buy green energy', 'buy more efficient car', 'switch electric car to car free', and switch to a 'plant-based diet'. Moderate-impact activities included 'replace gasoline with hybrid', 'wash clothes in cold water', 'recycle', and 'hang dry clothes'. The low-impact recommendation was to 'upgrade light bulbs' (p. 4). From their review of lifestyle changes promoted in government documents and textbooks, the researchers concluded that the tendency overall was to

focus on moderate- to low-impact actions, suggesting a missed opportunity which could be corrected in part with accurate information on the best choices.[5]

While it is both reasonable and legitimate to expect high-emitters to consider lifestyle changes, there are a number of issues worth highlighting in relation to studies such as this one, especially given the fact that it is precisely this type of study that gets picked up by national and international media outlets, helping to shape powerfully the public discourse around action, even in ways that the authors themselves may not have intended. Indeed, some of the headlines covering the study included: 'These Four Lifestyle Changes Will Do More to Combat Climate Change than *Anything Else*' (emphasis added), 'We're Teaching Our Kids the Wrong Ways to Fight Climate Change', and 'Here's How You Can Actually Help Stop Climate Change'. First, the approach fits neatly within dominant discourses that encourage us to focus on the individual instead of structures. Rather than pressuring governments to regulate so that there are limits to the proliferation of discount airlines that make flying so cheap it's hard to resist, or pressuring for reasonable limits on flying, approaches such as this emphasize personal lifestyle choices, running the risk that much more effective actions are crowded out, the possibility for politics diminished by an obsession with self. In this, rather than mobilizing politically to demand effective, accessible, and efficient public transportation for all, we focus on buying our hybrids or electric cars, or our green energy if we can afford it, revealing the elite politics that remains ingrained in so much action on climate change. While a focus on the individual does not preclude political mobilization, and it is fair to assume that the authors of the study might encourage political mobilization as well, we must pay close attention to how the discourse of the individual consumer and its constant iteration frame the discussion to such an extent that assuming the individual is the problem becomes common sense.

Second, this type of approach risks falling into the 'information deficit model of social action', which assumes that, with better knowledge and communication strategies that convey accurate information, individuals can then make the right choices.[6] While the authors do not endorse the model and acknowledge that it has been widely criticized, stating also that structural and cultural factors may work to impede lifestyle changes, they note nevertheless that their goal is 'to empower individuals to focus on changing the behaviours that are most effective at reducing their personal emissions' (Wynes et al., p. 5), drawing a connection, as such, between accurate information and empowerment. This connection, and how studies such as this tend to get covered, discourage a meaningful consideration of carbon's cultural politics. While we can ask whether or not an individual would choose one fewer transatlantic flight if given accurate information on the climate impact of that flight, we might also ask why we would necessarily mobilize politically to pressure governments to limit our ability as an individual to fly as much as we like. High-carbon lives are not simply about fulfilling structural imperatives; they simultaneously provide pleasure and comfort, while fulfilling individual 'desires'. This challenges us to think seriously, at the level of the individual, about what might be required to change our behaviours when high-carbon norms and cultural practices remain so deeply entrenched and so deeply fulfilling.[7]

Finally, studies such as this one fail to appreciate the extent to which the act of knowledge production and subsequent knowledge dissemination is itself deeply political. Can we assume that governments will always provide us with the best information? When in 2014 the Conservative government in Alberta began recruiting oil companies to help redesign its curriculum for students in Kindergarten to Grade 3, many were appalled, although perhaps not surprised. In a province whose economy depends heavily on the income it earns from the oil and gas sector, including oil companies in curriculum design made good economic sense. Education systems, along with many of our public institutions, are not value-neutral and free from political economic motivations, suggesting the need to consider how knowledge is constructed, by whom, and for what purposes.

What this discussion overall hopes to make clear is that dealing with climate change is fundamentally political, with political mobilization and collective action promising a much greater impact than going it alone. When we mobilize politically as citizens to demand regulatory frameworks that require system changes, the greater reach of these changes, while good for the climate, also promises greater social and ecological justice since the benefits of increased renewables capacity, more public transportation, more regulated energy efficiency, and so forth, are distributed with greater equity, rather than being contingent on whether or not you can personally afford to access them. In this, these are simultaneously struggles over democracy and democratic rights.

Yet how might the kinds of political mobilizations alluded to above be made possible? While there exists no meta-strategy in this respect, there are plenty of obstacles. Asking people to mobilize politically is difficult, with overlapping class, racialized, and gendered interests and power relations at play. Moreover, neoliberalism as a dominant ideology has been with us – to varying degrees, depending on one's location – since the 1980s, carrying with it a powerful ethos that prizes individual action and self-interest.

Social media distract further, with plenty of evidence now emerging to show how social media and smart phones in fact isolate us intensely, making it less likely that we will come together collectively, in person. Change appears much easier if it is the kind of change that one can control with much more immediate effects – carry your bag, shop differently, change your light bulbs. Indeed, what better way to counter the intense sense of powerlessness that sets in when confronted by the sheer magnitude of the challenge? It's relatively easy to think that, while you can't change the system, you can change yourself.

The work of scholar Massimo De Angelis is helpful here. In attempting to provide a theory of change, he notes that frameworks – and he speaks here of a particular variant of classical Marxism, whose theoretical orientation implies that system change must be total – i.e., from capitalism to socialism – can disempower, since the task seems too great and too far removed from the present. He suggests that, if the alternative to capitalism must be another '-ism', we risk not seeing the diverse movements that exist in the here and now and that are providing us with examples of alternatives – alternatives to capital and its drive to enclose and privatize, alternatives to livelihoods that prize the individual over the collective, and alternatives to fossil-fuelled and high-carbon lives dictated by powerful private interests.[8] What we see in these alternatives is that none succeeds by the will of someone acting alone – the consumer, the individual; rather, they are driven by collectives of people coming together as citizens to exercise their voice and political power. None is perfect or pure, and to expect either or both is to refuse to acknowledge the messiness and conflictual nature of movement building. Yet each can offer us lessons from which to learn, and from which we can draw in charting the best paths forward.

Lessons for the future

In his work on climate politics and social movement organizing amongst the poor in São Paulo, Brazil, Daniel Aldana Cohen distinguishes between what he calls 'luxury ecologies' and 'democratic ecologies'. With cities across the globe adopting policies aimed at cutting CO_2 emissions, including those that support densification over urban sprawl, the question of what kinds of ecologies will prevail is salient. For Cohen, 'luxury ecologies' are those prioritized in policies and projects that tend to benefit elite actors, including professional and business classes, especially in the real estate and financial sectors, while aligning to neoliberal modes of governance. 'Democratic ecologies', on the other hand, tend towards universalism, with the intent of supporting the poorest and most vulnerable. Between the two, we see a struggle between distinct variants of climate politics, between private and collective consumption, and private and collective rights. In other words, is it about private green spaces, high-priced high-rises, and the expansion of electric vehicle incentives, or is it about mass accessible public transit, affordable housing in a densified city, and community gardens? While the former is about lowering emissions in a way that benefits the few, the latter offers a low-carbon politics cut through with the everyday politics and struggles of the poor – for safe and affordable homes, for healthy food, for lives not spent commuting to and from illegal settlements on sprawling and congested roadways far from their places of work, for bodies not polluted by the highways and industries beside which they must live. In this, Cohen reminds us that every actor is an 'ecological actor', and every movement, an ecological movement, even if not conceived that way by its protagonists. All of this matters for the lessons it offers.[9]

In São Paulo, urban climate policies have oscillated between variations of luxury and democratic, with the municipal political landscape important in terms of government willingness to pursue one over the other; in this, it would seem that São Paulo's current millionaire populist mayor and his government will be more likely to advance the former over the latter. Cohen's is not a study on fairy-tale endings, therefore, but instead provides us with ways of thinking and doing climate politics more broadly, more democratically, and with greater justice. Under the right political conditions, São Paulo's various movements won 'partial victories', helping to pressure the Workers' Party to expand public transportation and infrastructure, to cancel fare hikes, to massively expand cycle lanes, to expropriate abandoned buildings from their private owners, and to plan 'denser, multiuse and multi-class corridors' (p. 155). The politics of São Paulo's social movements to advance the rights of the poor remind us not simply of the power of the collective to advance democratic citizenship, but also that the collective is essential to democratic ecologies. Equally important, Cohen illustrates the ways in which the concerns of the poor and working classes and those of the professional classes can intersect on the challenges of carbon – denser, multiuse, and multi-class corridors offer a broad range of benefits to groups with seemingly disparate interests, showing where spaces for collaboration might be possible.

The distinction between luxury and democratic ecologies can be extended further, both geographically and conceptually. Here we can return to the case studies of local or community control of renewables in Germany and Nova Scotia offered in chapter 4, adding the Canadian province of Ontario to the analysis. Recall that in Germany the history of movement organizing in support of environmental issues, combined with community experiments in renewables very early on, contributed to the emergence of political policies that facilitated community ownership of the country's sprawling renewables complex. This bottom-up support that helped create a broader movement with a stake in renewables in Germany has meant that, in the face of political attacks – some successful – the project overall has remained resilient. In Nova Scotia, on the other hand, the government's introduction of the Community Feed-in Tariff (COMFIT) in 2010, while requiring majority community control of renewables projects eligible under the policy – predominantly wind – and thus expanding the extent to which communities were involved in renewables deployment and ownership, was rooted nevertheless in a more top-down political process, absent a deep grassroots movement to sustain it. In the absence of deeper political support and a broad-based movement, the programme was vulnerable, going out with a whimper when cancelled by the Liberal government in 2016. In the province of Ontario, the Liberal government there introduced a Feed-in Tariff Program in 2009, with none of the community ownership requirements that were present in Nova Scotia. Instead, it was structured so that it favoured control and ownership by large commercial developers, to which the benefits would flow disproportionately as a result. These three examples allow us to look at luxury versus democratic ecologies from a different vantage point.

An ideal-type characterization of these three cases would see them as bottom-up (Germany), as mixed top-down/bottom-up (Nova Scotia), and as top-down (Ontario). Ontario's roll-out of wind energy was facilitated by the Green Energy Act in 2009. It was notable for shifting power to the province and away from local planning authorities, so that municipalities and communities lacked meaningful avenues to participate in the decision-making process. While there has been some community participation in the process, its technocratic structure in general favoured projects carried out by large corporate developers, some foreign-owned. It is the absence of both 'procedural' and 'distributive justice' in the process that has been identified as a significant source of opposition to wind energy in Ontario.[10] When the government cancelled the FIT in 2013, many celebrated. Importantly, a still-vocal movement opposing wind exists in Ontario, which might seriously complicate future plans for wind deployment in the province. On the other hand, while imperfect, Nova Scotia's COMFIT, designed to promote community ownership and benefits, meant that opposition was minimal, to almost non-existent in some locales, with municipalities and community groups engaged, to varying degrees, in the decision-making process around the development of projects. While there was insufficient time for deep roots to take hold and spread, on the whole Nova Scotia remains a more hospitable environment for contemplating the future expansion of wind energy. In sum, what we see in these three cases are competing ecologies – democratic and luxury. In Ontario, the government bypassed community and local participation to a significant extent, opting for a model within which CO_2 is rendered technical, the task of lowering it given disproportionately to private commercial interests whose profitability has little to do with the lives of those who must host the projects. Luxury here, then, concerns ownership and who benefits as a result of control and ownership structures. At the other end of the spectrum, community control and ownership of renewables projects in Germany, made possible in part by its history of social movement organizing, represents more democratic ecologies. Nova Scotia falls somewhere in between.

While the intent is not to romanticize the experience in Germany, or what it means to work with communities that are themselves differentiated and cut through with power, its experience helps to confirm the greater effectiveness of citizen mobilization in achieving more democratic, low-carbon ecologies, while simultaneously showing what more effective and democratic government policy can do.

We find more evidence of this in California amongst its well-organized environmental justice community. As a state, California exhibits severe levels of inequality and poverty. Data shows that low-income communities of colour are disproportionately impacted by environmental harm and toxic pollution due to their vulnerable socio-economic status, in combination with the fact that these communities live and work disproportionately in the most polluted areas of the state – next to freeways, refineries, heavily polluting industries, and toxic hotspots. For example, those living in California's 'toxic triangle' in the Oakland, Bayview–Hunters Point, and Richmond area die on average twelve years earlier than residents living in wealthier suburbs. They also suffer higher rates of cancer, respiratory disease, and other pollution-related illnesses, compounded by high unemployment, food insecurity, and poor public services.[11] These lived experiences engendered the growth of the state's powerful environmental justice (EJ) movement, organizing, working, and lobbying on behalf of these communities. When California announced plans to introduce its Cap-and-Trade Program, which took effect in 2012, many in the EJ community opposed it, since cap-and-trade allows industries to continue to pollute, trading or offsetting to meet their obligations under the Program. Ultimately unsuccessful in blocking cap-and-trade, an alliance of EJ organizations mobilized in turn to demand that carbon revenue earned from the sale of carbon allowances in state

auctions be used to benefit the communities they represented. Oganized into a potent political force able to lobby
effectively and work with various elected representatives,
the EJ community won partial victories on revenue spending. In particular, they were successful in having Senate
Bill 535 (SB 535) passed in 2012, requiring that 25 per cent
of auction revenue from the Greenhouse Gas Reduction
Fund support projects that benefit disadvantaged communities, with 10 per cent spent on projects located in
disadvantaged communities. The latter was increased to
25 per cent in 2016 with the passing of Assembly Bill 1550.
In order to identify disadvantaged communities throughout the state, the California Communities Environmental
Health Screening Tool, also known as CalEnviroScreen,
was created, using zip codes and pollution scores to map
their location. Since beginning, over $1.2 billion in carbon
revenue has gone to 327 projects benefitting disadvantaged
communities in California. Examples include the installa-
tion of home solar systems in Sacramento and Fresno in
1,600 and 1,212 low-income households, respectively. With
carbon revenue used to train workers from these communities, it was estimated that the systems would reduce
home energy bills by 75–90 per cent, with tens of thousands of dollars in savings per household. Energy-efficient
affordable housing units, low-cost public transportation,
green jobs, community green space and gardens, and
van-pooling for low-income agricultural workers are additional examples of how carbon revenue is being used for
low-carbon initiatives that provide benefits to low-income
communities.[12]

Struggles over how carbon revenue will be used can be
read through the lens of luxury versus democratic ecologies.
Historically, it has been the wealthy that have been able
to access renewables and clean technologies in California,

with many policies expanding benefits to these groups – electric car rebates is but one example. Without effective political mobilization, SB 535 would not exist, and there would be no state-mandated requirement to make sure that the poor and vulnerable are able to access, and benefit from, a low-carbon transition. Of course, this example illustrates simultaneously the partial, contingent, and often unsatisfactory nature of these struggles. The money itself comes from carbon trading, while future revenue spending and project funding is uncertain. In order to secure the recent extension of the Cap-and-Trade Program, the government of California agreed to include ACA 1, backed by Republicans and the oil industry, requiring a ballot measure in 2018 to amend the Constitution so that revenue spending after 2024 would require a two-thirds majority vote in both the Senate and the Legislature – a notoriously difficult requirement. This comes alongside other changes to the Program that would further diminish its chances of lowering emissions in the state.

Whether in São Paulo, Nova Scotia, Germany, or California, we are reminded that struggles for low-carbon democratic ecologies are never entirely settled – it's often one step forward, two steps back, depending on movement strength, political conditions, institutional configurations, and the power of those interests that have a deep stake in maintaining the current fossil order. Yet moving forward is much less likely in the absence of political mobilization on the part of citizens seeking to define future trajectories. The question of what kinds of ecologies will prevail depends deeply on who mobilizes and for what purpose. Take fossil fuel divestment campaigns, for example. Data suggests that in 2016 the value of investment funds – including pensions, investment firms, and sovereign wealth funds – globally committed to fossil fuel divestment topped $5.2

trillion. That is, investors worth $5.2 trillion are taking their money out of fossil fuels. This is happening for a number of reasons. One of the main driving factors for institutional investors has to do with the risk of stranded assets – assets that cannot pay off because they have lost their value in a low-carbon world. Fiduciary responsibility requires that investors assess risks and invest prudently on behalf of clients, and, in a rapidly warming world, fossil fuel assets are getting riskier by the day. If approached from the angle of stranded assets alone, however, divestment need not connect to democratic ecologies. Contemporary models of investment rely heavily on continued economic growth and development; if CO_2 is isolated, the goal remains one of putting money into high-growth sectors, as long as their carbon content is sufficiently low. Yet the fossil fuel divestment movement is in fact quite diverse, including student groups on university campuses, churches and faith-based organizations, pension funds, investors, foundations, and many more. For many of these groups, divestment is a moral and ethical issue, given the role played by fossil fuels in dangerously warming the planet, while devastating its most vulnerable inhabitants. Seen this way, fossil fuel divestment is deeply political, with the participation of diverse movements complicating the terrain upon which climate action takes place, achieving material successes that go beyond the simple logic of finance. What is ultimately clear is that only through citizen mobilization, the coming together of collectives with clear political goals and demands, are these outcomes possible.

Climate justice

This small blue planet of ours is the only home we've got. While the science on climate change and the damage we are doing to the world's ecosystems is solid beyond a shadow of a doubt, efforts to block a low-carbon transition, let alone one that promises democratic ecologies, remain troublingly powerful. The bitter irony of climate change is that those least responsible for it are already paying the biggest price, their homes washed away in storm surges and increasingly intense storms, their lands disappearing under the sea, their crops destroyed by crippling heat and drought, their stomachs empty for lack of food. In a world of over 7.5 billion people, there is no going back to some pristine mythical wilderness that only ever existed in our imaginations.[13] At the same time, continuing on the current path, with only minor tweaks, promises a dystopian future of the worst kind. What we need, therefore, are pathways that seek to redefine the 'good life' and what it means to live well. No longer can we ignore the contradictions and the anti-ecological tendencies embedded in an economic system that demands limitless growth while allowing obscene levels of wealth accumulation and growing inequality. But, as Naomi Klein reminds us, nor can we expect people to reject this system unless real alternatives exist, unless there are real options for a different kind of world. In this, it isn't enough to say 'no' – we need options for liveable and just futures.[14] Instead of providing a laundry list of movements – indeed, this chapter could have explored extensively the potentialities embedded in movements like La Via Campesina, bringing together farmers' organizations from around the world to fight for sustainable agriculture and food sovereignty, or those that are coming together in the Leap movement in Canada and that see indigenous communities, organized labour, and environmental groups working collectively – this chapter has focused on some of the broader lessons we must consider if we are to achieve successes, however partial they may be, in advancing democratic ecologies.

In this, it is perhaps fitting to end with the words of George Monbiot, a journalist for the *Guardian* newspaper out of the UK.[15] In a 2017 article published online, Monbiot asserts: 'without community, politics is dead'. Referencing our neoliberal times that have left us increasingly atomized and afraid, pursuing a 'politics of extreme individualism', he alludes to the fact that too many of us live lives without social relations that promote mutual understanding, deep cooperation, and sharing. While community in and of itself can be many things – dark and oppressive, or liberating and empowering – Monbiot's insights speak to one of the central lessons developed in this chapter: we cannot do this alone, as individuals or consumers. Indeed, it is only in the plural that we succeed in making demands to benefit all and not simply the few. It is only in the plural that we can force governments to stop subsidizing fossil fuels, to stop building pipelines, to invest in renewable energy, public services, public education, and green jobs. It remains uncertain how and whether or not we'll get there, but we must take the lessons where we find them and put them into practice. Those currently suffering under the devastating impacts of climate change, caused by too much carbon for far too long, the benefits of which have been enjoyed by far too few, are depending on it, as are the future generations yet to be born.

Notes

I THE PROBLEM OF CARBON

1 Christophe McGlade and Paul Ekins, 'The Geographical Distribution of Fossil Fuels Unused When Limiting Global Warming to 2°C', *Nature* 517: 7533 (2015): pp. 187–90; International Energy Agency, *2016 Key World Energy Statistics* (Paris: International Energy Agency, 2016); International Energy Agency, 'Energy Subsidies by Country', www.worldenergyoutlook.org/resources/energysubsidies; David Coady, Ian Parry, Louis Sears, and Baoping Shang, *How Large Are Global Energy Subsidies?* (Washington, DC: International Monetary Fund, 2015); T. F. Stocker, D. Qin, G.-K. Plattner, et al. (eds.), *Climate Change 2013: The Physical Science Basis. Contribution of Working Group I to the Fifth Assessment Report of the Intergovernmental Panel on Climate Change* (Cambridge University Press, 2013).
2 World Bank, 'CO_2 Emissions (Metric Tons Per Capita)', The World Bank Group, http://data.worldbank.org/indicator/EN.ATM.CO2E.PC; Carbon Disclosure Project, CDP *Carbon Majors Report 2017* (London: CDP Worldwide, 2017).
3 David Archer, *Global Warming: Understanding the Forecast*, 2nd edn (New York: John Wiley & Sons, 2012); David Bello, 'The Origin of Oxygen in the Earth's Atmosphere', *Scientific American* (2009), https://www.scientificamerican.com/article/origin-of-oxygen-in-atmosphere.
4 This section detailing the global carbon cycle and the formation of fossil fuels comes from the following sources: Stocker et al., *Climate Change 2013: The Physical Science Basis*, pp. 467–70; Scripps Institution of Oceanography, 'The Keeling Curve', UC San Diego, https://scripps.ucsd.edu/programs/

keelingcurve; US Department of Energy, 'How Fossil Fuels Were Formed', www.fe.doe.gov/education/energylessons/coal/gen_howformed.html.

5 Archer, *Global Warming: Understanding the Forecast*, pp. 29, 30–2.

6 Richard A. Betts, Chris D. Jones, Jeff R. Knight, Ralph F. Keeling, and John J. Kennedy, 'El Niño and a Record CO_2 Rise', *Nature Climate Change* (2016): pp. 806–10.

7 David Archer, Michael Eby, Victor Brovkin, et al., 'Atmospheric Lifetime of Fossil Fuel Carbon Dioxide', *Annual Review of Earth and Planetary Sciences* 37 (2009): pp. 117–34.

8 Kevin Anderson, 'Climate Change Going Beyond Dangerous: Brutal Numbers and Tenuous Hope', *Development Dialogue* 61 (2012): pp. 16–40 – emphasis in original.

9 Data in the following section on current climate emergencies is taken from the following sources: Jason Samenow, 'Two Middle East Locations Hit 129 Degrees, Hottest Ever in Eastern Hemisphere, Maybe World', *Washington Post* (2016), https://www.washingtonpost.com/news/capital-weather-gang/wp/2016/07/22/two-middle-east-locations-hit-129-degrees-hottest-ever-in-eastern-hemisphere-maybe-the-world/?tid=sm_tw; Drazen Jorgic and Syed Raza Hassan, 'Pakistan City Readies Graves, Hospitals, in Case Heat Wave Hits Again', Reuters (2016), https://www.reuters.com/article/pakistan-heatwave/pakistan-city-readies-graves-hospitals-in-case-heat-wave-hits-again-idUSL5N18D2T8; World Bank, *Turn Down the Heat: Why a 4 °C World Must Be Avoided* (Washington, DC: World Bank, 2012); Dale Kasler and Phillip Reese, 'California Drought Impact Pegged at $2.7 Billion', *Sacramento Bee* (2015), www.sacbee.com/news/state/california/water-and-drought/article31396805.html; 'Air Estimates Fort Mcmurray Wildfire Insured Losses at between $4.4 Billion and $9 Billion', *Canadian Underwriter* (2016), www.canadianunderwriter.ca/catastrophes/air-estimates-fort-mcmurray-wildfire-insured-losses-4-4-billion-9-billion-1004091472; United States Drought Monitor, 'Figure, Conditions for the Contiguous U.S.', National Drought Mitigation Center, http://droughtmonitor.unl.edu/Home/TabularStatistics.aspx; Suzanne Goldenberg, 'Western Antarctic Ice Sheet Collapse Has Already Begun, Scientist Warns', *The Guardian* (2014), https://www.theguardian.com/

environment/2014/may/12/western-antarctic-ice-sheet-collapse-has-already-begun-scientists-warn; Michael Slezak, 'Sections of Great Barrier Reef Suffering from "Complete Ecosystem Collapse"', *The Guardian* (2016) https://www.theguardian.com/environment/2016/jul/21/sections-of-great-barrier-reef-suffering-from-complete-ecosystem-collapse; Jean-Marie Robine, Siu Lan K. Cheung, Sophie Le Roy, et al., 'Death Toll Exceeded 70,000 in Europe During the Summer of 2003', *C. R. Biologies* 331: 2 (2008): pp. 171–8; Simon Albert et al., 'Interactions between Sea-Level Rise and Wave Exposure on Reef Island Dynamics in the Solomon Islands', *Environmental Research Letters* 11: 5 (2016); Malcolm McMillan, Amber Leeson, Andrew Shepherd, et al., 'A High-Resolution Record of Greenland Mass Balance', *Geophysical Research Letters* 43: 13 (2016): pp. 7002–10; World Wide Fund for Nature, 'Coral Reefs: Importance', http://wwf.panda.org/about_our_earth/blue_planet/coasts/coral_reefs/coral_importance.

10 Plan International, 'Typhoon Haiyan: Three Months On' (2014), https://reliefweb.int/report/philippines/typhoon-haiyan-three-months-o; Sabrina Shankman, 'Costs of Climate Change: Early Estimate for Hurricanes, Fires Reaches $300 Billion' (2017), https://insideclimatenews.org/news/28092017/ hurricane-maria-irma-harvey-wildfires-damage-cost-estimate-record-climate-change.

11 Andreas Malm, *Fossil Capital: The Rise of Steam Power and the Roots of Global Warming* (London: Verso, 2016).

12 Camilo Mora, Abby G. Frazier, Ryan J. Longman, et al., 'The Projected Timing of Climate Departure from Recent Variability', *Nature* 502: 7470 (2013): pp. 183, 185, 187.

13 Data on projected future climate emergencies comes from the following sources: Jeremy S. Pal and Elfatih A. B. Eltahir, 'Future Temperature in Southwest Asia Projected to Exceed a Threshold for Human Adaptability', *Nature Climate Change* 6: 2016 (2015): pp. 197–200; World Bank, *Turn Down the Heat: Confronting the New Climate Normal* (Washington, DC: World Bank, 2014), p. xix; Tord Kjellstrom, 'Impact of Climate Conditions on Occupational Health and Related Economic Losses: A New Feature of Global and Urban Health in the Context of Climate Change', *Asia-Pacific Journal of Public Health* 28: 2S (2015); Wolfram Schlenkera and Michael J. Roberts,

'Nonlinear Temperature Effects Indicate Severe Damages to U.S. Crop Yields under Climate Change', *PNAS* 106: 37 (2009): pp. 15594–8; World Bank, *Turn Down the Heat: Why a 4°C World Must Be Avoided*, p. xvii; Benjamin H. Strauss, Scott Kulp, and Anders Levermann, *Mapping Choices: Carbon, Climate, and Rising Seas, Our Global Legacy* (Princeton, NJ: Climate Central, 2015); Harry Verhoeven, 'Climate Change, Conflict and Development in Sudan: Global Neo-Malthusian Narratives and Local Power Struggles', *Development and Change* 42: 3 (2011): pp. 679–707; Markus G. Donat et al., 'More Extreme Precipitation in the World's Dry and Wet Regions', *Nature Climate Change* 6 (2016): pp. 508–13.

14 Michael Holden. 'Climate Change Root Cause of Syrian War: Britain's Prince Charles', Reuters (2015), https://www.reuters.com/article/us-climatechange-summit-charles/climate-change-root-cause-of-syrian-war-britains-prince-charles-idUSKBN0TC0N020151123.

15 Simon Dalby, 'The Environment as Geopolitical Threat: Reading Robert Kaplan's "Coming Anarchy"', *Cultural Geographies* 3: 4 (1996): pp. 472–96.

2 THE GLOBAL POLITICAL ECONOMY OF CARBON

1 Patrick T. Brown and Ken Caldeira, 'Greater Future Global Warming Inferred from Earth's Recent Energy Budget', *Nature* 552: 7684 (2017): pp. 45–50; Pep Canadell, Corinne Le Quéré, Glen Peters, Robbie Andrew, Rob Jackson, and Vanessa Haverd, 'Fossil Fuel Emissions Hit Record High After Unexpected Growth: Global Carbon Budget 2017', *The Conversation* (2016), https://theconversation.com/fossil-fuel-emissions-hit-record-high-after-unexpected-growth-global-carbon-budget-2017-87248; Zeke Hausfather, 'State of the Climate: 2017 Shaping up to be the Warmest Non-El Niño Year', *Carbon Brief* (9 Nov. 2016), https://www.carbonbrief.org/state-of-the-climate-2017-shaping-up-to-be-warmest-non-el-nino-year.

2 Karl Polanyi, *The Great Transformation: The Political and Economic Origins of Our Time* (Boston: Beacon, 1957).

3 Barry Bosworth, Gary Burtless, and Kan Zhang, *Later*

Retirement, Inequality in Old Age, and the Growing Gap in Longevity between the Rich and Poor (Washington, DC: The Brookings Institution, 2016).

4 Richard Peet, Paul Robbins, and Michael Watts, 'Global Nature', in Peet, Robbins, and Watts (eds.), *Global Political Ecology* (London: Routledge, 2011), pp. 1–48.

5 Malm, *Fossil Capital: The Rise of Steam Power and the Roots of Global Warming* (London:Verso, 2016).

6 IEA, *Global EV Outlook 2017 – 2 Million and Counting* (Paris: International Energy Agency, 2017), p. 6; Government of Canada, 'Greenhouse Gas Emissions by Economic Sector', https://www.canada.ca/en/environment-climate-change/services/environmental-indicators/greenhouse-gas-emissions/canadian-economic-sector.html; Union of Concerned Scientists, 'Car Emissions & Global Warming', https://www.ucsusa.org/clean-vehicles/car-emissions-and-global-warming#.WjMr4a2ZOCR; Statista, 'Number of Cars Sold Worldwide from 1990 to 2017 (in Million Units)', https://www.statista.com/statistics/200002/international-car-sales-since-1990.

7 IEA, *World Energy Outlook 2016 – Executive Summary* (Paris: International Energy Agency, 2016).

8 Details on the rise of coal-fired energy (steam) presented in this chapter are drawn from various sections of Malm's *Fossil Capital*. See the following pages specifically: pp. 36, 133–5, and 188–90.

9 Timothy Mitchell, *Carbon Democracy: Political Power in the Age of Oil* (London: Verso, 2013).

10 Elmar Altvater, 'The Social and Natural Environment of Fossil Capitalism', in Leo Panitch and Colin Leys (eds.), *Socialist Register 2007: Coming to Terms with Nature* (Monmouth: The Merlin Press, 2006), pp. 37–59.

11 Jutta Bolt, Marcel Timmer, and Jan Luiten van Zanden, 'GDP Per Capita since 1820', in Jan Luiten van Zanden, Joerg Baten, Marco Mira d'Ercole, Auke Rijpma, Conal Smith, and Marcel Timmer (eds.), *How Was Life? Global Well-Being since 1820* (Paris: OECD Publishing, 2014), pp. 65, 67. Note: GDP per capita is in US dollars at 1990 purchasing power parity. GDP is a rough measure of a country's economic wealth; in many cases, it is based on imprecise data, while it fails to tell us whether or not wealth is distributed equitably.

12 Amitav Ghosh, *The Great Derangement: Climate Change and the Unthinkable* (University of Chicago Press, 2016), pp. 106–7; David Harvey, *The New Imperialism* (Oxford University Press, 2003), pp. 42–9.

13 Zeke Hausfather, 'Mapped: The World's Largest CO_2 Importers and Exporters', *Carbon Brief* (5 July 2017), https://www.carbonbrief.org/mapped-worlds-largest-co2-importers-exporters.

14 See Karl Hallding, Marie Jürisoo, Marcus Carson, and Aaron Atteridge, 'Rising Powers: The Evolving Role of Basic Countries', *Climate Policy* 15: 5 (2013): pp. 608–31, for a discussion of emerging economies, carbon space, and development space.

15 Ibid.; Charlotte Streck and Maximilian Terhalle, 'The Changing Geopolitics of Climate Change', *Climate Policy* 13: 5 (2013): p. 533; Michael Grubb, 'Climate Policy: A New Era', *Climate Policy* 14: 3 (2014): p. 325.

16 Christine Shearer, Nicole Ghio, Lauri Myllyvirta, Aiqun Yu, and Ted Nace, *Boom and Bust 2016: Tracking the Global Coal Plant Pipeline* (CoalSwarm, Greenpeace, Sierra Club, 2016), pp. 3–4.

17 John Deutch, 'Decoupling Economic Growth and Carbon Emissions', *Joule* 1 (2017): pp. 3–9; IEA, 'Decoupling of Global Emissions and Economic Growth Confirmed', (2016), https://www.iea.org/newsroom/news/2016/march/decoupling-of-global-emissions-and-economic-growth-confirmed.html; Canadell et al., 'Fossil Fuel Emissions Hit Record High After Unexpected Growth'.

18 Gavin Bridge, 'Resource Geographies I: Making Carbon Economies, Old and New', *Progress in Human Geography* 35: 6 (2010): p. 821.

19 This section draws on the following sources: Robert Stavins, 'An Economic View of the Environment', www.robertstavinsblog.org; Robert Falkner, 'The Paris Agreement and the New Logic of International Climate Politics', *International Affairs* 92: 5 (2016): p. 1116; UNFCCC – Secretariat, *Paris Agreement*, Fccc/Cp/2015/10/Add.1 (Paris: United Nations Office at Geneva, 2015); Anna Petherick, 'Loss and Damage Post Paris', *Nature Climate Change* 6: August (2016): p. 741; Harriet Bulkeley and Peter Newell, *Governing Climate Change* (London: Routledge, 2010); Timmons Roberts and Romain Weikmans, 'The

Unfinished Agenda of the Paris Climate Talks: Finance to the Global South', *Planet Policy* (2015), https://www.brookings.edu/blog/planetpolicy/2015/12/22/the-unfinished-agenda-of-the-paris-climate-talks-finance-to-the-global-south. See Falkner, 'The Paris Agreement', pp. 114–18, for a detailed summary of the key features of the Agreement.

20 See, for example, Matthew J. Hoffmann, *Climate Governance at the Crossroads: Experimenting with a Global Response after Kyoto* (Oxford University Press, 2011); Karin Bäckstrand, Jonathan W. Kuyper, Björn-Ola Linnér, and Eva Lövbrand, 'Non-State Actors in Global Climate Governance: From Copenhagen to Paris and Beyond', *Environmental Politics* 26: 4 (2017): pp. 561–79. It is Bäckstrand et al. who introduce the concept of 'hybrid multilateralism' in their analysis of the evolving global climate governance regime that is discussed above.

3 TRADING CARBON TO COOL THE WORLD?

1 Population Matters, 'PopOffsets: Smaller Families, Less Carbon', www.popoffsets.org; the Cheat Neutral video can be viewed here: https://www.youtube.com/watch?v=f3_CYdYDDpk.

2 George Orwell, *Nineteen Eighty-Four* (London: Penguin Books Ltd, 1949), pp. 241–2.

3 Robert Cox, 'Social Forces, States and World Orders: Beyond International Relations Theory,' *Millennium Journal of International Studies* 10: 2 (1981).

4 World Bank, 'Carbon Pricing Dashboard', http://carbonpricingdashboard.worldbank.org; High-Level Commission on Carbon Prices, 'Report of the High-Level Commission on Carbon Prices' (Carbon Pricing Leadership Coalition, 2017), p. 1.

5 This section draws on the following sources: Ronald Coase, 'The Problem of Social Cost', *Journal of Law & Economics* 3 (1960): pp. 1–44; Arno Simons and Jan-Peter Voß, 'Politics by Other Means: The Making of the Emissions Trading Instrument as a 'Pre-History' of Carbon Trading', in Benjamin Stephan and Richard Lane (eds.), *The Politics of Carbon Markets* (London: Routledge, 2015), pp. 51–68; Richard Schmalensee and Robert

N. Stavins, 'The SO$_2$ Allowance Trading System: The Ironic History of a Grand Policy Experiment', *MIT Center for Energy and Environmental Policy Research Working Paper* August 2012: CEEPR WP 2012 – 012 (2012), pp. 1–18; Jonas Meckling, *Carbon Coalitions: Business, Climate Politics, and the Rise of Emissions Trading* (Cambridge, MA: MIT Press, 2011); Peter Newell and Matthew Paterson, *Climate Capitalism: Global Warming and the Transformation of the Global Economy* (New York: Cambridge University Press, 2010); UNFCCC, '1997 Kyoto Protocol to the UN Framework Convention on Climate Change (UNFCCC)', UN Doc FCCC/CP/1997/Add.1, Dec. 10, 1997; 37 ILM 22 (1997); World Bank and Ecofys, *Carbon Pricing Watch 2017* (Washington, DC: World Bank, 2017), p. 7; Peter Newell, 'The Political Economy of Global Environmental Governance', *Review of International Studies* 2008: 34 (2008): pp. 507–29; UNFCCC – Secretariat, *Paris Agreement*, Fccc/Cp/2015/10/Add.1 (Paris: United Nations Office at Geneva, 2015); Matthew Paterson, 'Who and What Are Carbon Markets For? Politics and the Development of Climate Policy', *Climate Policy* 12: 1 (2012): pp. 82–97.

6 Naomi Oreskes and Erik M. Conway, *Merchants of Doubt: How a Handful of Scientists Obscured the Truth on Issues from Tobacco Smoke to Global Warming* (New York: Bloomsbury, 2010).

7 Paterson, 'Who and What Are Carbon Markets For?'

8 Ibid.

9 World Bank and Ecofys, *Carbon Pricing Watch 2017*, p. 7; World Bank, 'Carbon Pricing Dashboard'.

10 Maria Gutierrez, 'Making Markets out of Thin Air: A Case of Capital Involution', *Antipode* 43: 3 (2011): pp. 639–61; Donald MacKenzie, 'Making Things the Same: Gases, Emission Rights and the Politics of Carbon', *Accounting, Organizations and Society* 34: 3–4 (2009): pp. 440–55; Larry Lohmann, 'The Endless Algebra of Climate Markets', *Capitalism, Nature, Socialism* 22: 4 (2011): pp. 93–116.

11 Robert W. Hahn and Robert N. Stavins, 'The Effect of Allowance Allocations on Cap-and-Trade System Performance', *Journal of Law and Economics* 54: 4 (2011): pp. 267–94; Blas Pérez Henríquez, *Environmental Commodities Markets and Emissions Trading* (New York: Resources for the Future Press, 2013).

12 Tania M. Li, 'Rendering Society Technical: Government through Community and the Ethnographic Turn at the World Bank in Indonesia', in David Mosse (ed.), *Adventures in Aidland: The Anthropology of Professionals in International Development* (Oxford: Berghahn, 2011), pp. 57–80.

13 Data on GWPs is taken from the following sources: G. Myhre, D. Shindell, F.-M. Bréon et al., '2013: Anthropogenic and Natural Radiative Forcing', in T. F. Stocker, D. Qin, G.-K. Plattner, et al. (eds.), *Climate Change 2013: The Physical Science Basis. Contribution of Working Group I to the Fifth Assessment Report of the Intergovernmental Panel on Climate Change* (Cambridge University Press, 2013); Paulo Artaxo, Terie Bernsten, Richard Betts, et al., 'Changes in Atmospheric Constituents and Radiative Forcing', in S. Solomon, D. Qin, M. Manning, et al. (eds.), *Climate Change 2007: The Physical Science Basis. Contribution of Working Group I to the Fourth Assessment Report of the Intergovernmental Panel on Climate Change* (Cambridge University Press, 2007); UNEP DTU Partnership, 'CDM Projects by Type', UNEP DTU Partnership, Centre on Energy, Climate and Sustainable Development, http://cdmpipeline.org/cdm-projects-type.htm.

14 Myhre et al., '2013: Anthropogenic and Natural Radiative Forcing', p. 711.

15 See: Carbon Market Watch, 'Coal Power Projects in the CDM', http://carbonmarketwatch.org/category/coal-power; 'CDM Investors Snub Coal Power', *Carbon Market Watch Newsletter* (September 2013): p. 4.

16 The remainder of this section and its discussion of uncertainty in calculating baselines and counterfactuals in the process of commensuration is drawn from the following sources: Larry Lohmann, 'Marketing and Making Carbon Dumps: Commodification, Calculation and Counterfactuals in Climate Change Mitigation', *Science as Culture* 14: 3 (2005): pp. 203–35; Adam Bumpus, 'The Matter of Carbon: Understanding the Materiality of Tco2e in Carbon Offsets', *Antipode* 43: 3 (2011): pp. 612–38; Gregory L. Simon, Adam Bumpus, and Philip Mann, 'Win-Win Scenarios at the Climate–Development Interface: Challenges and Opportunities for Stove Replacement Programs through Carbon Finance', *Global Environmental Change* 22: 2012 (2012): pp. 275–87; Gutierrez, 'Making Markets out of Thin Air';

Cathleen Fogel, 'The Local, the Global, and the Kyoto Protocol', in S. Jasanoff and M. Martello (eds.), *Earthly Politics: Local and Global in Environmental Governance* (Cambridge, MA: The MIT Press, 2004), pp. 103–26; Coline Seyller, Sebastien Desbureaux, Symphorien Ongolo, et al., 'The "Virtual Economy" of REDD+ Projects: Does Private Certification of REDD+ Projects Ensure Their Environmental Integrity?' *International Forestry Review* 18 (2016): pp. 231–46; Kate Dooley, *Misleading Numbers: The Case for Separating Land and Fossil Based Carbon Emissions*, ed. FERN (Brussels: FERN, 2014).

17 Morgan Robertson, 'Discovering Price in All the Wrong Places: The Work of Commodity Definition and Price under Neoliberal Environmental Policy', *Antipode* 39: 3 (2007): pp. 500–26.

18 Ingo Tschach, 'Flexible Auction Mechanisms', in ICIS (ed.), *Carbon Markets Almanac 2014: Global Developments and Outlook* (London: ICIS, 2014), pp. 8–9.

19 Lesley K. McAllister, 'The Overallocation Problem in Cap-and-Trade: Moving Toward Stringency', *Columbia Journal of Environmental Law* 34: 2 (2009): pp. 395–445, 410; Stefan E. Weishaar, *Emissions Trading Design: A Critical Overview* (Cheltenham: Edward Elgar, 2014), pp. 100–4; RGGI Inc., 'RGGI States Propose Lowering Regional CO_2 Emissions Cap 45%, Implementing a More Flexible Cost-Control Mechanism' (New York: RGGI, 2013); Judith Schroeter, 'Regional Greenhouse Gas Initiative (RGGI)', in ICIS (ed.), *Carbon Markets Almanac 2015: Global Developments & Outlook* (London: Reed Business Information Ltd, 2015), pp. 90–2; Peter Shattuck and Jordan Stutt, *The Regional Greenhouse Gas Initiative: A Model Program for the Power Sector* (Boston: Acadia Center, 2015), p. 4; Potomac Economics, *Market Monitor Report for Auction 12* (Fairfax, VA: Potomac Economics, 2011); Dan X. McGraw, 'Weekly Review: CCAs Decline as Market Takes Bearish View on Auction', in *ICIS Tschach Solutions* (ICIS Tshach Solutions, 2015); Lesley K. McAllister, 'Auction Results in California Cap and Trade', Environmental Law Prof Blog, http://lawprofessors.typepad.com/environmental_law/2013/05/auction-results-in-california-cap-and-trade-.html; Simon Chen, 'China Pilot Emissions Trading Schemes', in ICIS (ed.), *Carbon Markets Almanac 2015*, pp. 48–63; Clayton Munnings, Richard Morgenstern, Zhongmin Wang, and Xu Liu, *Assessing the Design*

of *Three Pilot Programs for Carbon Trading in China* (Washington, DC: Resources for the Future, 2014); Stian Reklev, 'Decisions, Decisions: Key Issues China Must Solve for Its National Carbon Market', *Carbon Pulse* (2015), http://carbon-pulse.com/7348.

20 Larry Lohmann, 'Uncertainty Markets and Carbon Markets: Variations on Polanyian Themes', *New Political Economy* 15: 2 (2010): p. 243.

21 European Commission, 'The EU Emissions Trading System', http://ec.europa.eu/clima/policies/ets/faq_en.htm; Matt Gray and Bryony Worthington, *The ETS in Context: Understanding and Managing EU ETS Policy Interactions* (London: Sandbag, 2015); Damien Meadows, Yvon Slingenberg, and Peter Zapfe, 'EU ETS: Pricing Carbon to Drive Cost-Effective Reductions across Europe', in Jos Delbeke and Peter Vis (eds.), *EU Climate Policy Explained* (London: Routledge, 2015), pp. 29–60; Emily Spears, 'New Zealand Emissions Trading Scheme', in ICIS (ed.) *Carbon Markets Almanac 2015*, pp. 70–3.

22 Carbon Market Watch, *Industry Windfall Profits from Europe's Carbon Market: How Energy-Intensive Companies Cashed in on Their Pollution at Taxpayers' Expense* (Brussels: Carbon Market Watch, 2016); *Carbon Leakage Myth Buster* (Brussels: Carbon Market Watch, 2015); Sandbag, *Discharging a Political Storm: Supporting EU Competitiveness and Innovation in the ETS – Interactive Data Tool*, https://sandbag.org.uk/blog/2015/jul/13/discharging-political-storm-supporting-eu-competit.

23 Martin Cames, Ralph O. Harthan, Jurg Fussler, et al., *How Additional Is the Clean Development Mechanism? Analysis of the Application of Current Tools and Proposed Alternatives* (Berlin: Prepared for DG CLIMA, 2016).

24 Anja Kollmuss, Lambert Schneider, and Vladyslav Zhezherin, 'Has Joint Implementation Reduced Ghg Emissions? Lessons Learned for the Design of Carbon Market Mechanisms', in *SEI Policy Brief* (Seattle, WA: Stockholm Environment Institute, 2015).

4   CARBON TRANSITIONS

1 Slavoj Žižek, *Violence* (New York: Picador, 2008).

2 The concept of pathways can be found in much of the literature

on low-carbon transitions and green futures. See the following
for examples that help to systematize possible pathways, their
challenges, and their implications: Ian Scoones, Melissa Leach,
and Peter Newell, 'The Politics of Green Transformations',
in Scoones, Leach, and Newell (eds.), *The Politics of Green
Transformations* (London: Routledge, 2015), pp. 1–24; Jennifer
Clapp and Peter Dauvergne, *Paths to a Green World: The Political
Economy of the Global Environment* (Cambridge, MA: MIT Press,
2011).

3 L. Frank Baum, *The Wonderful Wizard of Oz* (New York:
   Penguin, 2006).

4 Global CCS Institute, *The Global Status of CCS: 2016 – Summary
   Report* (Melbourne, 2016).

5 Glen P. Peters, Robbie M. Andrew, Josep G. Canadell, et al.,
   'Key Indicators to Track Current Progress and Future Ambition
   of the Paris Agreement', *Nature Climate Change* 7 (2017):
   p. 121.

6 Vivian Scott, Stuart R. Haszeldine, Simon F. B. Tett, and
   Andreas Oschlies, 'Fossil Fuels in a Trillion Tonne World',
   Nature Climate Change 5 (2015): p. 422.

7 Ian Austin, 'Technology to Make Clean Energy from Coal
   Is Stumbling in Practice', *New York Times* (29 March 2016),
   https://www.nytimes.com/2016/03/30/business/energy-
   environment/technology-to-make-clean-energy-from-coal-is-
   stumbling-in-practice.html?_r=0.

8 The data and quotations on NETs in this section are taken from:
   Pete Smith, Steven J. Davis, Felix Creutzig, et al., 'Biophysical
   and Economic Limits to Negative CO_2 Emissions', *Nature
   Climate Change* 6 (2016): pp. 42–50; Kevin Anderson and Glen
   Peters, 'The Trouble with Negative Emissions', *Science* 354:
   6309 (2016): pp. 182–3.

9 Saturnino M. Borras Jr and Jennifer C. Franco, 'Global Land
   Grabbing and Trajectories of Agrarian Change: A Preliminary
   Analysis', *Journal of Agrarian Change* 12: 1 (2012): pp. 34–59;
   Rachel A. Nalepa and Dana Marie Bauer, 'Marginal Lands:
   The Role of Remote Sensing in Constructing Landscapes for
   Agrofuel Development', *Journal of Peasant Studies* 39: 2 (2012):
   pp. 403–22.

10 The data covered in the next two paragraphs comes from the
   following sources: David P. Keller, Ellias Y. Feng, and Andreas

Oschlies, 'Potential Climate Engineering Effectiveness and Side Effects During a High Carbon Dioxide-Emission Scenario', *Nature Communications* 5: 3304 (2014): pp. 1–11; Victor Brovkin, Vladimir Petoukhov, Martin Claussen, Eva Bauer, David Archer, and Carlo Jaeger, 'Geoengineering Climate by Stratospheric Sulfur Injections: Earth System Vulnerability to Technological Failure', *Climatic Change* 92 (2009): pp. 243–59; Jim M. Haywood, Andy Jones, Nicolas Bellouin, and David Stephenson, 'Asymmetric Forcing from Stratospheric Aerosols Impacts Sahelian Rainfall', *Nature Climate Change* 3 (2013): pp. 660–5.

11 The information and data in the following three paragraphs come from the following sources: Jasmin Cantzler, Andrzei Ancygier, Fabio Sferra, et al., 'Foot Off the Gas: Increased Reliance on Natural Gas in the Power Sector Risks an Emissions Lock-In', Briefing Paper for Climate Action Tracker (2016); Ed King, 'Oil Majors' Climate Plan gets Hostile Reception', *Climate Home* (2016), www.climatechangenews.com /2016/11/04/oil-majors-climate-plan-gets-hostile-reception.

12 Xiaochun Zhang, Nathan P. Myhrvold, Zeke Hausfather, and Ken Caldeira, 'Climate Benefits of Natural Gas as a Bridge Fuel and Potential Delay of Near-Zero Energy System', *Applied Energy* 167 (2016): pp. 317–22.

13 Deyi Hou, Jian Luo, and Abir Al-Tabbaa, 'Shale Gas Can Be a Double-Edged Sword for Climate Change', *Nature Climate Change* 2 (2012): pp. 385–7; Alireza Babaie Mahani, Ryan Schultz, Honn Kao, Dan Walker, Jeff Johnson, and Carlos Salas, 'Fluid Injection and Seismic Activity in the Northern Montney Play, British Columbia, Canada, with Special Reference to the 17 August 2015 Mw 4.6 Induced Earthquake', *Bulletin of the Seismological Society of America* 107 (2017): pp. 542–52.

14 Scoones et al., 'The Politics of Green Transformations'; Frank W. Geels, Benjamin K. Sovacool, Tim Schwanen, and Steve Sorrell, 'The Socio-Technical Dynamics of Low-Carbon Transitions', *Joule* 1: 3 (2017): pp. 463–79.

15 Renewable energy data for this chapter is taken from the following sources: International Energy Agency, *Renewables Information (2016 Edition)* (Paris: International Energy Agency, 2016); Frankfurt School et al., *Global Trends in Renewable*

Energy Investment 2017 (Frankfurt School of Finance &
Management, 2017); IRENA (International Renewable Energy
Agency), 'Renewable Energy Highlights', www.irena.org/
DocumentDownloads/Publications/IRENA_Renewable_
energy_highlights_July_2017.pdf.

16 Mark Z. Jacobson and Mark A. Delucchi, 'Providing All
Global Energy with Wind, Water, and Solar Power, Part I:
Technologies, Energy Resources, Quantities and Areas of
Infrastructure, and Materials', *Energy Policy* 39: 3 (2011): pp.
1154–69; David Schwartzman, 'How Much and What Kind of
Energy Does Humanity Need?' *Socialism and Democracy* 30: 2
(2016): pp. 97–120.

17 Mariana Mazzucato, 'The Green Entrepreneurial State', and
Stephen Spratt, 'Financing Green Transformations', in Scoones
et al. (eds.), *The Politics of Green Transformations*.

18 Tadzio Mueller, *Diversity Is Strength: The German Energiewende
as a Resilient Alternative* (London: The New Economics
Foundation, 2017); Colin Nolden, 'Governing Community
Energy – Feed-in Tariffs and the Development of Community
Wind Energy Schemes in the United Kingdom and Germany',
Energy Policy 63 (2013): pp. 543–52.

19 Stephen Thomas, 'Lessons from the Community Feed-in
Tariff (COMFIT)', presentation to the 3rd meeting of SECURE
Project Partners, 'Transnational Good Practices and Knowledge
Transfer', 18 May 2017, Solleftȧ, Sweden.

20 Sandra van Niekerk, 'Public Renewable Energy in Africa: The
Potential for Democratic Electrification', in David A. McDonald
(ed.), *Making Public in a Privatized World* (London: Zed Books,
2016); David Ockwell and Rob Byrne, *Sustainable Energy for
All: Innovation, Technology, and Pro-Poor Green Transformations*
(London: Routledge, 2017), p. 145.

21 Daniel Chavez and Satoko Kishimoto, 'Towards Energy
Democracy: Discussions and Outcomes from an International
Workshop', *Workshop Report* (Amsterdam: Transnational
Institute, 2016); Daniel Chavez, 'The Coopelesca Cooperative
Experience: Energy Democracy at Work in a Rural Context',
Energy Democracy, www.energy-democracy.net/?p=367.

22 Marcus Power, Peter Newell, Lucy Baker, Harriet Bulkeley,
Joshua Kirshner, and Adrian Smith, 'The Political Economy
of Energy Transitions in Mozambique and South Africa: The

Role of the Rising Powers', *Energy Research and Social Science* 17 (2016): pp. 10–19.

5 THE FUTURE OF CARBON POLITICS

1 Sian Sullivan, 'What's Ontology Got to Do with It? On Nature and Knowledge in a Political Ecology of the "Green Economy"', *Journal of Political Ecology* 24 (2017): pp. 217–42; Marion Zimmer Bradley, *The Mists of Avalon* (New York: Ballantine Publishing Group, 1982).
2 World Bank, *World Development Indicators 2008* (Washington, DC: World Bank, 2008), p. 4.
3 Bruno Latour, *We Have Never Been Modern* (Cambridge, MA: Harvard University Press, 1993).
4 Chris Edwards and Jonna Meyhoff Fry, 'Life Cycle Assessment of Supermarket Carrier Bags: A Review of the Bags Available in 2006' (Bristol: Environment Agency, 2011).
5 Seth Wynes and Kimberly A. Nicholas, 'The Climate Mitigation Gap: Education and Government Recommendations Miss the Most Effective Individual Actions', *Environmental Research Letters* 12 (2017): pp. 1–9.
6 Harriet Bulkeley, Matthew Paterson, and Johannes Stripple, *Towards a Cultural Politics of Climate Change: Devices, Desires and Dissent* (Cambridge University Press, 2016), p. 5.
7 Ibid.
8 Massimo De Angelis, 'Separating the Doing and the Deed: Capital and the Continuous Character of Enclosures', *Historical Materialism* 12: 2 (2004): pp. 57–87.
9 Daniel Aldana Cohen, 'The Other Low-Carbon Protagonists: Poor People's Movements and Climate Politics in São Paulo', in Miriam Greenberg and Penny Lewis (eds.), *The City Is the Factory: New Solidarities and Spatial Strategies in an Urban Age* (Ithaca, NY: Cornell University Press, 2017), pp. 143, 155–6.
10 Chad Walker and Jamie Baxter, 'Procedural Justice in Canadian Wind Energy Development: A Comparison of Community-Based and Technocratic Settings', *Energy Research and Social Science* 29 (2017): pp. 160–9.
11 Vien Truong, 'Addressing Poverty and Pollution: California's

Sb 535 Greenhouse Gas Reduction Fund', *Harvard Civil Rights – Civil Liberties Law Review* 29: 2 (2014): p. 495.

12 Alvaro S. Sanchez, *California's Climate Investments: 10 Case Studies – Reducing Poverty and Pollution* (Berkley, CA: The Greenlining Institute, 2015).

13 William Cronon, 'The Trouble with Wilderness: Or, Getting Back to the Wrong Nature', in W. Cronon. (ed.), *Uncommon Ground: Toward Reinventing Nature* (New York: W. W. Norton, 1995), pp. 69–90.

14 Naomi Klein, *This Changes Everything: Capitalism vs. the Climate* (New York: Simon and Schuster, 2014).

15 George Monbiot, 'This Is How People Can Truly Take Back Control: From the Bottom Up', *The Guardian* (8 February 2017), https://www.theguardian.com/commentisfree/2017/feb/08/take-back-control-bottom-up-communities.

Selected Readings

For those readers interested in learning more about the science of carbon dioxide and climate change, a number of books by David Archer are particularly useful, including *The Global Carbon Cycle* (Princeton University Press, 2010), *The Long Thaw: How Humans Are Changing the Next 100,000 Years of Earth's Climate* (Princeton University Press, 2009), and *Global Warming: Understanding the Forecast*, 2nd edn (New York: John Wiley & Sons, 2012). James Hansen's *Storms of My Grandchildren: The Truth About the Coming Climate Catastrophe and Our Last Chance to Save Humanity* (New York: Bloomsbury, 2010) and Michael E. Mann and Lee R. Kump's *Dire Predictions: Understanding Climate Change* (New York: Dorling Kindersley, 2008) also help to unpack the science, with the latter providing an illustrated guide to understanding the research coming out of the Intergovernmental Panel on Climate Change. Very much connected to the science of climate change is the issue of climate change denial. A must-read on the history and politics of climate change denial in the United States is Naomi Oreskes and Erik M. Conway's *Merchants of Doubt: How a Handful of Scientists Obscured the Truth on Issues from Tobacco Smoke to Global Warming* (New York: Bloomsbury, 2010). Michael E. Mann's *The Hockey Stick and the Climate Wars: Dispatches from the Front Lines* (New York: Columbia University Press, 2012) and Mann and Tom Toles' *The Madhouse Effect: How Climate Change Denial is Threatening*

Our Planet, Destroying Our Politics, and Driving Us Crazy
(New York: Columbia University Press, 2010) provide further insight into the political battles that are being waged to
undermine the science of climate change.

There are many books that examine the history of
fossil fuel use and development, their relationship to
the current global political economy, and the role they
are playing in global climate change. Some excellent
sources to consider include Gavin Bridge and Phillipe Le
Billon's *Oil* (Cambridge: Polity, 2017), also part of Polity's
Resources series; Timothy Mitchell's *Carbon Democracy:
Political Power in the Age of Oil* (London: Verso, 2013);
Suzana Sawyer's *Crude Chronicles: Indigenous Politics,
Multinational Oil, and Neoliberalism in Ecuador* (London:
Duke University Press, 2004); and Hannah Appel, Arthur
Mason, and Michael Watts' edited book *Subterranean
Estates: Life Worlds of Oil and Gas* (Ithaca, NY: Cornell
University Press, 2015). Other important contributions that
examine the political economy of climate change and capi-
talism include Tim DiMuzio's *Carbon Capitalism: Energy,
Social Reproduction, and World Order* (London: Rowman
& Littlefield, 2015); Andreas Malm's *Fossil Capitalism:
The Rise of Steam Power and the Roots of Global Warming*
(London: Verso, 2016); Jason W. Moore's edited volume
*Anthropocene or Capitalocene? Nature, History, and the Crisis
of Capitalism* (Oakland,CA: PM Press, 2016); Peter Newell's
*Globalization and the Environment: Capitalism, Ecology,
and Power* (Cambridge: Polity, 2012); Adrian Parr's *The
Wrath of Capital: Neoliberalism and Climate Politics* (New
York: Columbia University Press, 2013); and Christopher
Wright and Daniel Nyberg's *Climate Change, Capitalism,
and Corporations: Processes of Creative Self-Destruction*
(Cambridge University Press, 2015).

The literature on carbon and climate change governance

is now extensive, considering the question of governance from a range of perspectives, angles, scales, and sites. For the reader interested in understanding the increasingly diverse networks and sites within which governance happens – public, private, and from the local to the global – excellent resources are Harriet Bulkeley, Liliana B. Andonova, Michelle M. Betsill, et al.'s *Transnational Climate Change Governance* (New York: Cambridge University Press, 2014); Harriet Bulkeley's *Accomplishing Climate Governance* (New York: Cambridge University Press, 2016); Harriet Bulkeley and Peter Newell's *Governing Climate Change* (London: Routledge, 2010); and Craig A. Johnson's forthcoming book *The Power of Cities in Global Climate Change Politics: Saviours, Supplicants or Agents of Change?* (Basingstoke: Palgrave MacMillan, 2018). For those interested in exploring the politics of climate change adaptation, Marcus Taylor's *The Political Ecology of Climate Change Adaptation: Livelihoods, Agrarian Change and the Conflicts of Development* (New York: Routledge, 2015) provides an insightful analysis. On the theme of climate change and development more generally, including its impacts on vulnerable populations and how they are responding, Hannah Reid's *Climate Change and Human Development* (London: Zed Books, 2014) is particularly useful.

Readers wanting to advance further their understanding of carbon trading and carbon offsetting can look to Steffen Böhm and Siddartha Dahbi's edited volume *Upsetting the Offset: The Political Economy of Carbon Markets* (London: Mayfly Books, 2009); Anthony Hall's *Forests and Climate Change: The Social Dimensions of REDD in Latin America* (Cheltenham: Edward Elgar, 2012); Janelle Knox-Hayes' *The Cultures of Carbon Markets: The Political Economy of Climate Governance* (Oxford University Press, 2016); Melissa Leach and Ian Scoones' edited book *Carbon Conflicts and Forest*

Landscapes in Africa (London: Routledge, 2015); Larry Lohmann's 'Uncertainty Markets and Carbon Markets: Variations on Polanyian Themes' (*New Political Economy* 15: 2 (2010): pp. 225–54), and his work on 'Commodity Fetishism in Climate Science and Policy' (www.the cornerhouse.org.uk/sites/thecornerhouse.org.uk/files/ Fetishism.pdf); Jonas Meckling's *Carbon Coalitions: Business, Climate Politics, and the Rise of Emissions Trading* (Cambridge, MA: MIT Press, 2011); Peter Newell and Matthew Paterson's *Climate Capitalism: Global Warming and the Transformation of the Global Economy* (New York: Cambridge University Press, 2010); Stephanie Paladino and Shirley J. Fiske's edited book *The Carbon Fix: Forest Carbon, Social Justice, and Environmental Governance* (London: Routledge, 2017); and Benjamin Stephan and Richard Lane's edited volume *The Politics of Carbon Markets* (London: Routledge, 2015). The subscription news service Carbon Pulse is an invaluable resource for those hoping to stay on top of the rapidly evolving world of global carbon markets and related developments in the field of climate policy across the globe (http://carbon-pulse.com).

Literature dealing with carbon transitions is also especially diverse. Some good places to start include Steffen Böhm, Zareen Pervez Bharucha, and Jules Pretty's edited book *Ecocultures: Blueprints for Sustainable Communities* (London: Routledge, 2015); Patrick Bond's *Politics of Climate Justice: Paralysis Above, Movement Below* (Scottsville, South Africa: University of KwaZulu-Natal Press, 2012); Lester R. Brown's *The Great Transition: Shifting from Fossil Fuels to Solar and Wind Energy* (New York: W. W. Norton & Company, 2015); Naomi Klein's *This Changes Everything: Capitalism vs. the Climate* (New York: Simon and Schuster, 2014) and *No Is Not Enough: Resisting the New Shock Politics and Winning the World We Need* (Toronto: Knopf

Canada, 2017); Craig Morris and Arne Jungjohann's *Energy Democracy: Germany's Energiewende to Renewables* (London: Palgrave MacMillan, 2016); David Ockwell and Rob Byrne's *Sustainable Energy for All: Innovation, Technology, and Pro-Poor Green Transformations* (London: Routledge, 2017); and Ian Scoones, Melissa Leach, and Peter Newell's edited book *The Politics of Green Transformations* (London: Routledge, 2015).

Finally, for those interested in exploring further the issue of consumption, environmentalism, cultural politics, and the role of the individual in dealing with climate change, excellent resources are Harriet Bulkeley, Matthew Paterson, and Johannes Stripple's edited book *Towards a Cultural Politics of Climate Change: Devices, Desires and Dissent* (Cambridge University Press, 2016); Peter Dauvergne's *Environmentalism of the Rich* (Cambridge, MA: MIT Press, 2016) and *The Shadows of Consumption: Consequences for the Global Environment* (Cambridge, MA: MIT Press, 2008); David McDermott Hughes' *Energy Without Conscience: Oil, Climate Change, and Complicity* (London: Duke University Press, 2017); Thomas Hylland Eriksen's *Overheating: An Anthropology of Accelerated Change* (London: Pluto Press, 2016); Kari Marie Norgard's *Living in Denial: Climate Change, Emotions, and Everyday Life* (Cambridge, MA: MIT Press, 2011); Thomas Princen, Michael Maniates, and Ken Conca's *Confronting Consumption* (Cambridge, MA: MIT Press, 2002); and Heather Rogers' *Green Gone Wrong: Dispatches from the Frontlines of Eco-Capitalism* (London: Verso, 2010).

Index